Implementation of
ROBOT SYSTEMS

Implementation of
ROBOT SYSTEMS

An introduction to robotics, automation, and successful systems integration in manufacturing

MIKE WILSON

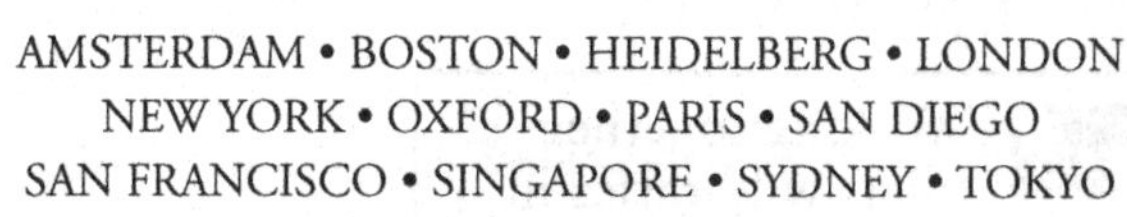

Butterworth-Heinemann is an imprint of Elsevier
32 Jamestown Road, London NW1 7BY, UK
525 B Street, Suite 1800, San Diego, CA 92101-4495, USA
225 Wyman Street, Waltham, MA 02451, USA
The Boulevard, Langford Lane, Kidlington, Oxford OX5 1GB, UK

First published 2015

British Library Cataloguing in Publication Data

A catalogue record for this book is available from the British Library

Library of Congress Cataloging-in-Publication Data

A catalog record for this book is available from the Library of Congress

ISBN: 978-0-124-04733-4

For information on all Butterworth-Heinemann publications visit our website at **store.elsevier.com**

Printed and bound in the United States

CONTENTS

ACKNOWLEDGEMENTS

I would like to thank Elsevier for the opportunity to put into print the knowledge and experience I have gained over the 30 years I have worked in robotics. In particular thank you to Hayley Gray and Charlie Kent of Elsevier, who provided much needed encouragement during the more challenging phases of this project. Thanks also to Brian Wilson, my father, who set me on the road into engineering and has encouraged me at all stages of my life, in addition to providing advice on this work.

I am grateful to all my friends and the colleagues with whom I have worked throughout my career. It has been a pleasure to meet and work with so many people from different countries and industries with a shared interest in automation and robotics.

Finally, thank you to my wife Elena, for her patience and support, throughout the writing of this book.

DEDICATION

This book is dedicated with love, to my wife Elena and my three daughters, Rosie, Robyn, and Emily.

ABOUT THE AUTHOR

Mike Wilson has worked in the robotics industry for over 30 years. He qualified with a masters degree in Industrial Robotics from Cranfield University in 1982.

His initial experience was within the British Leyland car company working on the development and implementation of robot systems, particularly for adhesive, sealant, and paint applications. In 1988, he moved into sales, initially with Torsteknik (which ultimately became part of Yaskawa), selling robotic welding systems to a range of automotive component and metal fabrication businesses in the UK. This was followed by a move to GMF (which became Fanuc Robotics), where he initially concentrated on the automotive sector followed by general sales management, finally becoming UK managing director, responsible for all aspects of the business including sales, finance, engineering, and customer service.

This was followed by 6 years with Meta Vision Systems, a venture capital-backed UK business focussed on vision guidance systems for robots and welding machines. This period included the acquisition and subsequent integration of two competitors, one based in Montreal and the other in the UK. Over 95% of Meta's business was outside the UK, which resulted in many visits to overseas customers, particularly throughout Europe, India, and North America.

In 2005, Mike started his own business providing consultancy services to manufacturing companies and automation suppliers, as well as training. This included projects for Italian, Korean, Dutch, and UK companies, retention as an expert witness for a number of disputes, as well as teaching on behalf of Warwick University. In 2012, Mike joined ABB Robotics in the UK in a sales management role.

Throughout his career, Mike has also been very active in trade associations and other related organisations in the UK. He has been involved with the British Automation and Robot Association since 1991, serving as chairman since 2009. He has also been the chairman of the International Federation of Robotics for the period 2000–2003, the only chairman to be elected for two consecutive terms.

ABOUT THE AUTHOR

LIST OF FIGURES

LIST OF TABLES

CHAPTER 1

Introduction

Chapter Contents

Abstract

This chapter outlines the contents of the book and provides a brief history of automation, differentiating between process and discrete automation. To this end, the chapter surveys the history of industrial robots from the first installation in the 1960s onward, outlining the key milestones in the development of industrial robot technology. The chapter also discusses the development of robot applications, particularly those driven by the automotive industry, as well as the effects of robot use on employment.

Keywords: Industrial robots, Discrete automation, Factory automation, Unimation, PUMA, Robot density

The advent of industrial robots in the 1960s heralded an exciting period for manufacturing engineers. These machines provided them an opportunity to automate activities in ways that had previously been infeasible. In 1961, General Motors first applied an industrial robot in a manufacturing process. Since that time, robotic technology has developed at a fast pace, and today's robots are very different from the first machines in terms of performance, capability, and cost. Over 2 million robots have been installed across many industrial sectors, and a whole new automation sector has developed. These robots have provided significant benefits to manufacturing businesses and consumers alike. There are many challenges involved in achieving successful applications, however, and over the last 50 years, those who have led the way have learnt many lessons.

The challenges are largely caused by the limitations of robots in comparison with humans. Although they can perform many manufacturing tasks as well as, or even better than, humans, robots do not presently have the same sensing capabilities and intelligence as humans do. Therefore, to achieve a successful application, these limitations have to be considered, and the application must be designed to allow the robot to perform the task successfully.

This book provides a practical guide for engineers and students hoping to achieve successful robot implementation. It is not intended to provide exhaustive details of robot technology or how robots operate or are programmed. It is intended to convey lessons learnt from experience, offering guidance particularly to those who are new to the application of robots. The fear of problems and unfulfilled expectations is often the largest barrier to the introduction of robots. Even given the current population of robots, many companies throughout the world can still benefit from adopting this technology. Their reticence to incorporate robotics is largely due to a fear of the unknown, a view that robots are "fine for the automotive industry but they are not for us". This mistaken view holds back the growth and profitability of many companies that have not embraced robot technology nor gained the benefits it can bring.

1.1 SCOPE

As mentioned above, this book is intended to be a guide to the practical application of robot systems. Many academic books describe the development and current technologies of robotics. Many examples of applications are also supplied by robot manufacturers and system integrators via the internet. Yet, few sources cover all the important aspects of the implementation of robot systems. Many experts have developed this knowledge through experience, but most have not had the time to impart this experience to others in this way.

In the following pages, we introduce automation. Knowledge of automation varies across different industry sectors. Therefore, it is important to understand when robots are appropriate and, most importantly, when they are not. The term *robot* also conjures up many different images from simple handling devices to intelligent humanoid machines. So, we provide an explanation for the term *industrial robot*, which then defines the context for this book.

Although we do not intend to provide a deep understanding of robot technology, we do offer an introduction to the benefits of using robots, as well as robot configurations, performance, and characteristics. This knowledge is required as a starting point for all applications because it serves as the basis for selecting a suitable robot for a particular application. This is covered in Chapter 2.

A robot consists of a mechanical device, typically an arm and its associated controller. On its own, this device can achieve nothing. In order to perform an application, a robot must be built into a system that includes many other devices. Chapter 3 provides a brief outline of the most important equipment that can be used around a robot.

Chapter 4 then reviews typical applications. Again, we do not intend this review to be exhaustive. Instead it provides examples of a range of robot applications throughout various different industry sectors. These are used to illustrate the main issues that must be addressed when implementing a robot solution, particularly those issues relevant to a specific sector or application.

The remainder of the book outlines a step-by-step process that can be followed in order to achieve a successful application. First, in Chapter 5, we discuss the initial process of developing the solution, although the process is normally iterative, with the actual solution often not finalised until the financial justification has been developed. A key element of any successful implementation is the definition of the system specification. In most cases, a company subcontracts the actual implementation of the robot solution to an external supplier, such as a system integrator, and this supplier must have a clear understanding of both the requirements for the system and the constraints under which it is to perform. These are defined in the User Requirements Specification. Without this specification, the chance of failure is greatly increased due to the lack of a clear understanding between the customer and supplier. The purpose of the user requirements specification is to convey this information, and we discuss the development of this key document in Chapter 6.

Of course, the implementation of a robot system must provide benefits to the end user. These benefits are often financial, and the financial justification must be clearly identified at the commencement of the project. Normally, a company will not proceed with the purchase of a robot system, as with most other capital investments, unless the financial justification is viable. For this reason the final decision maker, within the end user, requires a compelling financial justification. Therefore, the development of this justification is as important as the engineering design of the solution. This is not just a case of determining labour savings. Robot systems also provide many other benefits that can be quantified financially. In many cases, robot systems are not implemented, because the justification does not satisfy the financial requirements of the business. However, a detailed analysis presented in the correct way can improve the justification. This is covered in Chapter 7.

All successful projects require a methodical approach to project planning and management. In this respect, robot systems implementation is no different, although specific issues must be addressed, particularly for those companies undertaking an initial implementation of robot technology. Chapter 8 provides a guide to the successful implementation of a robot system from the initial project plan, through supplier selection to the installation and operation of the robot system. In particular, the chapter considers common problems and how they can be avoided.

Finally, Chapter 9 summarises the implementation process. This chapter also provides some thoughts as to how engineers and companies that are new to robot technologies might benefit from the development of an automation strategy. This strategy offers a plan from which manufacturers can develop their expertise and automation use as part of the overall company goals.

1.2 INTRODUCTION TO AUTOMATION

Automation can be defined as "automatically controlled operation of an apparatus, process, or system by mechanical or electronic devices that take the place of human labour". Basically, automation is the replacement of man by machine for the performance of tasks, and it can provide movement, data gathering, and decision making. Automation therefore covers a very wide array of devices, machines, and systems ranging from simple pick-and-place operations to the complex monitoring and control systems used for nuclear power plants.

Industrial automation originated with the Industrial Revolution and the invention of the steam engine by James Watt in 1769. This was followed by the Jacquard punch card-controlled loom in 1801 and the cam-programmable lathe in 1830. These early industrial machines can be more appropriately defined as mechanisation because they were exclusively mechanical devices with little programmability. In 1908, Henry Ford introduced mass production with the Model T, and Morris Motors in the UK further enhanced this process in 1923 by employing the automatic transfer machine. The first truly programmable devices did not appear until the 1950s, with the development of the numerically controlled machine tool at MIT. General Motors installed the first industrial robot in 1961 and the first programmable logic controller in 1969. The first industrial network, the Manufacturing Automation Protocol was conceived in 1985, and all of these developments have led to the automation systems in use today.

Robots are a particular form of automation. To understand the role robots can play within a manufacturing facility, one must distinguish between the different types of automation. The first major distinction is between process and discrete automation. Discrete, or factory, automation provides the rapid execution of intermittent movements. This frequently involves the highly dynamic motion of large machine parts that must be moved and positioned with great precision. The overall production plant generally consists of numbers of machines from different manufacturers that are often independently automated. In contrast, process automation is designed for continuous processes. The plant normally consists of closed systems of pumps used to move media through pipes and valves connecting containers in which materials are added and mixing and temperature control takes place. In simple terms, discrete automation is normally associated with individual parts, whereas process automation controls fluids.

The control systems for chemical plants and oil refineries provide examples of process automation. The facilities used by the automotive industry represent discrete automation, and some facilities in the food and beverage sector require both forms of automation. In these facilities, process automation provides the basic product (such as milk), and factory automation then provides the handling when the product has been put into discrete packages, the bottles or cartons.

Therefore, robots are a form of discrete or factory automation. Within this group the types of automation can be categorised as hard or soft automation. Hard automation is dedicated to a specific task, and, as a result, it is highly optimised to the performance of that task. It has little flexibility but can operate at very high speeds. An excellent example of hard automation is cigarette-manufacturing machinery. Soft automation provides flexibility. It can either handle different types of product through the same equipment or be reprogrammed to perform different tasks or operate on different products. The trade-off is often performance, in that soft automation is not as optimised, and therefore, it cannot achieve the same output as dedicated, hard automation. Robots are a very flexible form of soft automation because the basic robot can be applied to many different types of application.

1.3 EVOLUTION OF ROBOTS

The word "robot" was first used by the Czech playwright Karel Capek in his play "Rossum's Universal Robots". It is derived from the Czech word "robota" meaning slave labour. This science fiction play, from 1920, portrayed robots as intelligent machines serving their human masters but ultimately taking over the world. The popular concept of robots has emerged from this beginning. Other writers developed the ideas further. In particular, in the 1940s, Isaac Asmiov created three laws of robotics to govern the operation of his fictional robots (Engelberger, 1980):

1. Robots must not injure humans, or through inaction, allow a human being to come to harm.
2. Robots must obey the orders given by human beings except where such orders would conflict with the first law.
3. Robots must protect their own existence as long as this does not conflict with the first or second law.

Although these laws are fictional, they do provide the basis used by many current researchers developing robot intelligence and human–robot interaction.

Robots come in many forms. Due to the high profile of fictional robots, such as C3PO from *Star Wars*, the public often associates robots with intelligent, humanoid devices, but the reality of current robot technologies is very different. The robot community categorises robots into two distinct application areas, service robots and industrial robots. Service robots are being developed for a wide range of applications, including unmanned aircraft for the military, machines for milking cows, robot surgeons, search and rescue robots, robot vacuum cleaners, and educational and toy robots. Due to the wide range of applications and environments in which they operate, the machines vary greatly in terms of size, performance, technology, and cost. The use of service robots is a growing market, largely addressed by companies other than those that supply the industrial sector. There is some cross over in terms of technologies with industrial robots, but the machines are very different.

This book focuses on the use of robots in the industrial sector. These machines have been developed to meet the needs of industry and therefore they have much less variation than do service robots. The following is an accepted definition (ISO 8373) for an industrial robot (International Federation of Robotics, 2013).

> *An automatically controlled, re-programmable, multipurpose manipulator programmable in three or more axes, which may be either fixed in place or mobile for use in industrial automation applications.*

This provides a distinction between robots and other automation devices such as pick-and-place units, machine tools, and storage-and-retrieval systems.

The industrial robot industry began in 1956 with the formation of Unimation by Joseph Engelberger and George Devol. Devol had previously registered the patent "Programmed Article Transfer", and together, they developed the first industrial robot, the Unimate (Figure 1.1). Unimation installed the first robot into industry for the stacking of die cast parts at the General Motors plant in Trenton, New Jersey. This robot was a hydraulically driven arm that followed step-by-step instructions stored on a magnetic drum. The first major installation was again at General Motors, in this

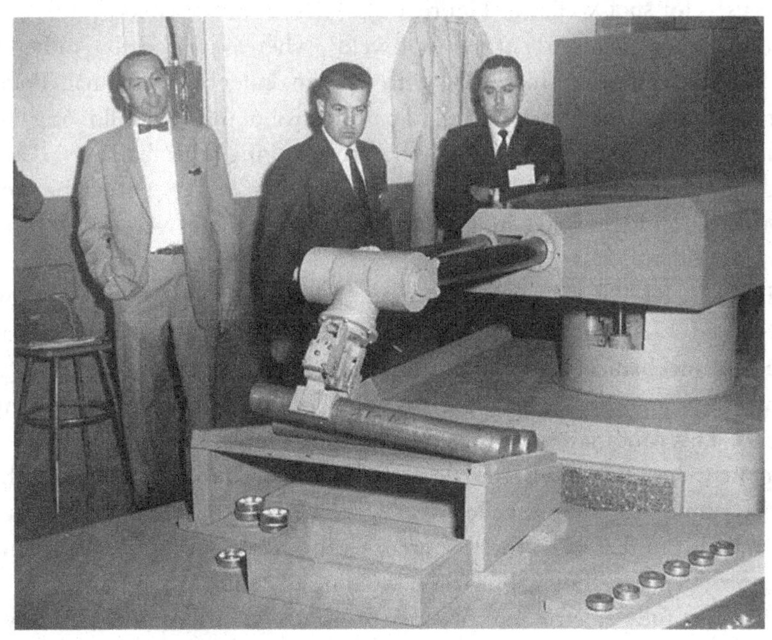

Figure 1.1 First Unimate.

Figure 1.2 General Motors, Lordstown robot installation.

case, at the Lordstown assembly plant in 1969, where Unimation robots were used for spot welding (Figure 1.2). These robots enabled automation to address more than 90% of the spot welds, whereas, previously, only 40% had been processed automatically, with the remainder being manual. Trallfa, Norway, offered the first commercial painting robot in 1969, following their earlier development for in-house use, spray-painting wheelbarrows. Robot production then commenced in Japan, following an agreement between Unimation and Kawasaki in 1969, and by 1973, there were 3000 robots in use worldwide.

In 1973, KUKA, Germany, developed their own robots, having previously used Unimation machines. These robots were the first to have six electromechanical driven axes. Also in that year, Hitachi became the first company to incorporate vision sensors to allow the robot to track moving objects. This robot fastened bolts on a moving mould for the production of concrete piles. By 1974, the first commercially available robot with a minicomputer-based controller was available, the Cincinnati Milacron T3, and also in that year, the first arc welding robots were installed. These were produced by Kawasaki for the welding of motor cycle frames.

The first fully electric microprocessor-controlled robot, the IRB 6, was launched by ASEA in Sweden in 1974. This machine mimicked the human

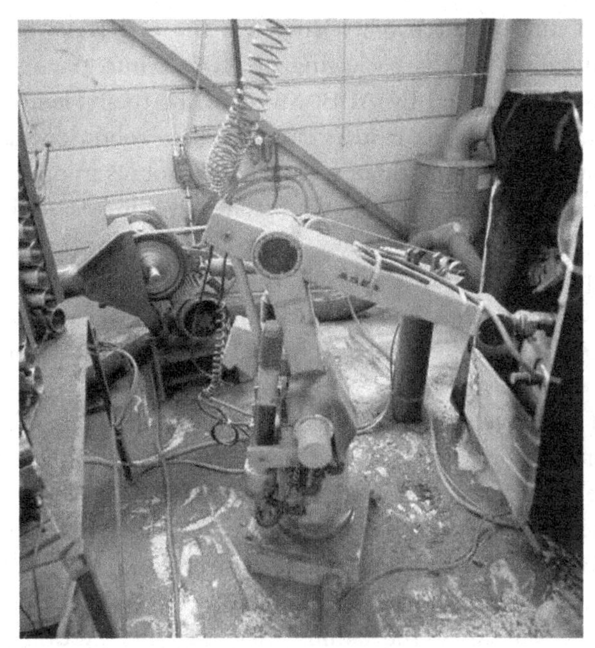

Figure 1.3 First IRB 6 installation.

arm and had a carrying capacity of 6 kg. The first unit was installed for the polishing of stainless steel tubes (Figure 1.3). In 1975, the Olivetti SIGMA robot, based on a Cartesian design (see Section 2.1), was one of the first to be used for assembly applications, and in 1978, Unimation, with support from General Motors, developed the programmable universal machine for assembly (PUMA). Also in 1978, the first selective compliance assembly robot arm (SCARA) (see Section 2.1) was developed. Both the PUMA and SCARA designs were developed for assembly applications with limited carrying capacity but good repeatability and high speed.

In 1979, Nachi, Japan developed the first heavy-duty electrically driven robots. These machines provided improved performance and reliability over the hydraulically driven robots, and electric drives became the industry standard. In 1981, the first gantry robot was introduced, providing a much larger range of motion than could be achieved with conventional designs. By 1983, 66,000 robots were in operation.

The first direct-drive SCARA robots were launched in 1984 by Adept, USA. The motors were directly connected to the arms, eliminating the need for intermediate gears, chains or belts. The simplified design provided high accuracy and improved reliability. In 1992, the first Delta configuration robot was installed for the packaging of pretzels into trays. This design was subsequently adopted by ABB, Sweden, for their FlexPicker robot, which, at the time, was the world's fastest picking robot, operating at 120 picks per minute. By 2003, there were 800,000 robots in operation.

By 2004, the capability of robot controllers had increased significantly, and Motoman, Japan, launched a new controller that provided synchronised control of up to 38 axes and could control four robot arms simultaneously. Other controller developments included the use of PC-based systems running Windows CE and touch screens and colour displays for the teach pendants.

Many of the developments resulted from the needs of the automotive industry. Since the first installation at General Motors, the automotive sector has been the major user of industrial robots, and it remains the largest customer for the robot industry. However, today's robots are also used in a very wide range of other sectors, from food to aerospace. The International Federation of Robotics (IFR) has been collecting robot installation data since 1988, and it publishes the annual statistics in "World Robotics" (International Federation of Robotics, 2013). The widespread use of robots across many industry sectors is illustrated in Figure 1.4. Robots are also in use in every industrialised country, and Figure 1.5 illustrates the growth of robot applications across the world.

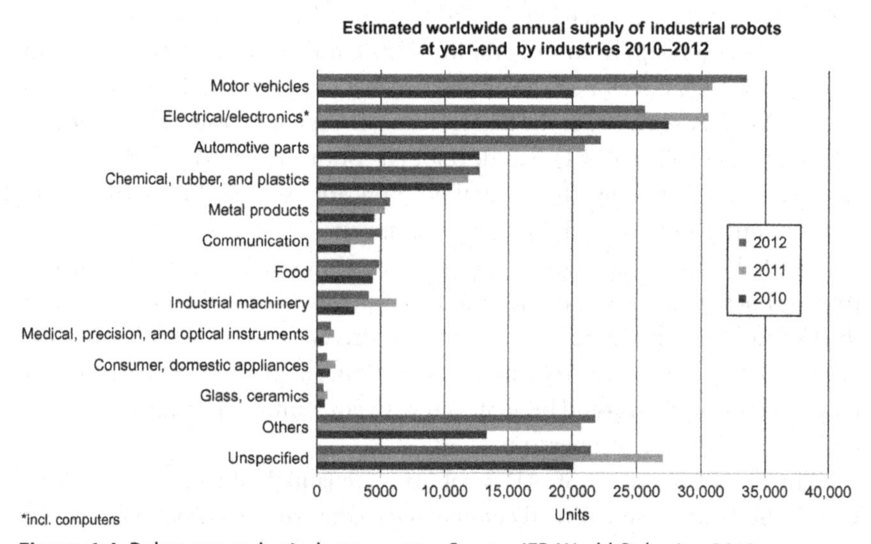

Figure 1.4 Robot usage by industry sector. *Source: IFR World Robotics, 2013.*

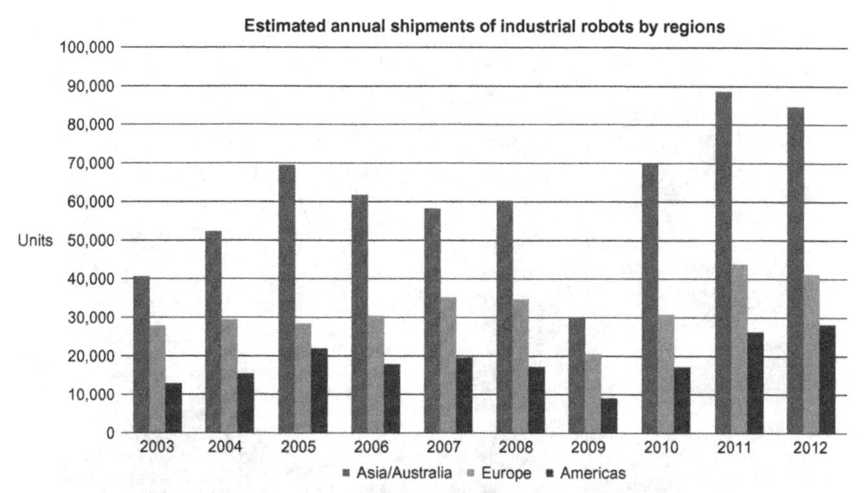

Figure 1.5 Worldwide robot usage. *Source: IFR World Robotics, 2013.*

1.4 DEVELOPMENT OF ROBOT APPLICATIONS

There is a very broad range of applications for robots across all sectors of manufacturing, as well as in other sectors, including the film and entertainment industries. The IFR report "World Robotics" states that, by the end of 2012, over 1 million robots were to be in operation (International Federation of Robotics, 2013). Of these, close to 50% were estimated to be in use in the automotive industry, including automotive components. The sector with the second largest number of robotic applications was the electrical and electronics industry with about 18%, which was followed by the plastics and chemicals industries at 10% and the metals and machinery sector with 9%. The automotive sector has always been the leading user of robots, and as a result, it has had a major impact not only on the development of robots but also the development of their applications.

1.4.1 Automotive Industry

The initial applications in the automotive industry were mainly spot welding to form the car body from the individual pressings. The repeatability of the robots, combined with their dexterity, allowed the automation of the large numbers of manual welding stations. These applications were developed from the manual approach, and therefore, they tended to include transfer devices moving the parts between welding stations, with a number of robots

Figure 1.6 Spot welding in a body shop.

manipulating welding guns at each station (Figure 1.6). The spot welding equipment was based on the manual packages with remote transformers and heavy cables providing the power to the welding guns. The early robots were hydraulic to provide the carrying capacity required, although these were quickly replaced by electric servo-driven machines because the electric machines provided improved performance and reliability. As the weight carrying capacity of the electric robots increased, transformer guns were developed in order to integrate the transformer within the weld gun, which removed the need for the heavy-duty cables. These advances increased the reliability of the operation because the cables did not wear as quickly as those used previously.

More recently, robots have been used to handle the parts, particularly for the subassemblies, through the weld cell, using either fixed guns or more robots with weld guns. This has provided enhanced flexibility, and it has also increased the number of robots. The drivers for this new approach have been the shorter life span for car designs and also the larger number of variants being produced at a single facility.

The technology and performance of the welding equipment has also been improved. For example, some plants have introduced servo-driven weld guns, with the operation of the gun being fully controlled by the robot. These machines provide the benefit of reduced cycle times because the closing and opening of the gun can be initiated before the robot reaches the required positions and the size of the opening can be controlled for specific sequences of spot welds. Similar operations, such as self-pierce riveting and stud welding, have also been automated in large numbers using robots. In addition, industry has shown increasing interest in laser welding, with some systems in use. These tend to be more expensive systems, but the performance and results justify the increased cost in some cases. Dedicated laser welding robots have been developed to provide integrated solutions for the feed of the laser to the head, increasing reliability as well as performance.

Painting and underbody sealing are other applications that were introduced at an early stage. These were driven by the unpleasant nature of the applications and the need to achieve consistent quality. Paint applications initially used hydraulic robots because the environment of the paint booth, with high concentrations of solvent, would not allow electric drives. The development of explosion proof paint robots solved this problem and significantly improved the performance and ease of use of painting robots. The initial robots used a lead teach principle, where the painter would actually hold the robot arm and take it through the required path, often whilst painting. The robot would record this path and then repeat it. The merit of this approach is that it allowed paths to be created quickly, but these robots had limited capability for path editing. Therefore, the development of complex programmes could be very time-consuming. Off-line programming, using a simulation of the robot, booth, and car body, quickly became the preferred approach and this approach is now very effective and widely used.

Within a typical spray booth, a number of other automated devices are used for painting, including reciprocators carrying spray guns and also electrostatic bell machines. These are very appropriate for the coverage of the exterior surfaces, if the line is continuously moving. Robots were often used in addition to these machines to provide the coverage for the interiors, including the engine bays, boot and door shuts. To provide access to these areas, opening devices were developed to operate in conjunction with the paint robots. Also servo-driven tracks allowed the robot to follow the car body as it moved through the booth.

Today, automotive robot paint systems are fully integrated process solutions, with closed-loop control of the paint and air services. This provides the ability to adjust the painting parameters during a path to provide the optimum results. In addition, colour changers are integrated within the robot arm to minimise the time and paint wastage during colour changes. Robots can also be equipped with a range of paint applicators, including air spray, electrostatic air spray, and electrostatic bells. Each of these has specific advantages, and therefore, the robot can be fitted with the optimum equipment for a particular application.

Underbody sealing was an unpleasant manual application for which robots were applied. In the early days, paint robots were utilised due to the ease of programming. However, these applications are now normally achieved using standard robots. Robots also addressed seam sealing due to the need for precise application of the sealer to the joints between the panels on the car body. This required the use of standard robots to achieve the precision required. Vision systems have been used to identify the position of the car body to ensure that the system can accommodate variations in this position and maintain the precise application of the sealant bead. Over time the applicators were improved to include the use of closed-loop flow control to provide more precise control of the application, as well as the ability to vary the output of the applicator to suit the seams to be addressed. Adhesive applications were also quickly introduced, again driven by quality. These included direct glazing, applying the adhesive and then fitting the glass to the car body, and various applications through which adhesive is applied to panels prior to assembly, as with the bonnet inner and outer panels.

The panels for the car bodies are produced within press shops. These may be within the automotive original equipment manufacturer (OEM) or one of their tier 1 suppliers. Robots have been used on press lines for many years to provide the transfer of panels between presses. These robots replaced either manual transfer or the more dedicated "iron hand" and other mechanical devices. In some cases, the press lines have been developed further, and now large transfer presses include their own handling equipment, with robots being used to load and unload these machines. However, there is still a place for the more traditional press line, which often uses robots between the presses.

The introduction of robots to the final assembly operations was slower than it was in the body shops largely because of the difficulties in handling the parts involved and the need to work inside the car body. However, many applications now have the potential to be automated using robots, and a company's decision as to the route (e.g., manual, semi-automated or fully automated using robots) is made on the basis of cost rather than technical barriers.

Similarly, the uptake of robots in engine assembly operations was initially slow. These were generally high-volume and highly automated facilities that used more dedicated equipment. As the need for flexibility has increased, the benefit of using robots has led to significant use within modern engine assembly facilities. In addition to general handling and the transfer of parts between machines, applications include assembly and deburring.

1.4.2 Automotive Components

The automotive components sector has followed the automotive industry in the use of robots. They have embraced robotic technology as a means to achieving the quality and flexibility required by their customers, the automotive OEMs. The applications across the sector are more varied due to the range of different parts being produced. In most cases, the robot cells tend to be standalone rather than elements of a large automation system, because of the smaller number of operations required to produce a completed assembly. The applications generally fall into a number of categories: plastics, metal working, electrics and electronics, and assembly.

There are many plastic parts required for a vehicle, including both interior parts, such as dash panels and trim, and exterior parts, such as bumpers, door handles, and spoilers. The robot applications used to produce these parts include injection mould machine unloading, routing, water jet cutting, assembly (including bonding and welding) and painting.

The metal working applications include the production of subassemblies for the vehicle body, as well as other large items such as the exhaust system. In terms of subassemblies for the car body, such as suspension mountings, the applications include press loading and unloading applications, as well as spot and arc welding. Arc welding is the main application of robots in exhaust system manufacturing, during which the machines assemble various parts of the exhaust system, such as pipes, flanges, mounting brackets, silencer boxes, and catalytic converters. It is worth noting the majority of arc welding for automotive parts is performed by the component suppliers rather than in the main car plants.

There are many mechanical parts that ultimately become part of the powertrain. Applications here include machine tool loading and unloading, grinding, deburring and other metal finishing applications. Similarly, the lights, air conditioning units, electrics, and other subassemblies within a car are assembled and tested using a variety of robot applications.

1.4.3 Other Sectors

The electrical and electronics industries have seen strong growth in the use of robots over recent years. Many dedicated machines are currently in use, for example, populating printed circuit boards. However, robots are used for machine loading and unloading, testing, assembling larger components and many other applications. The growth has been largely driven by the increased production of consumer electronic devices such as mobile phones and tablets, with the majority of this activity taking place in Asia.

The food industry has been seen as a large potential user for robots for many years, mainly due to the high number of manual operations involved in this sector. However, there has not yet been a widespread take off in the number of robots used, and a number of major challenges have yet to be fully addressed. Firstly, the cost of labour is lower than it is in other sectors, such as automotive, making the justification of automation more difficult. The products are very often organic and can be difficult to handle, in addition to being inconsistent in size and shape. Hygiene rules, particularly for operations on naked food products, also require specific standards for the automation equipment, such as wash-down, which again increases cost. The ultimate goal would be to achieve a fully automated food processing plant that provides much improved hygiene, because the main source of contaminants, the workforce, is removed.

1.4.4 Future Potential

One of the analyses performed by the IFR is the calculation of robot density, or the number of robots per 10,000 employees in manufacturing industry. Figure 1.7 shows a comparison between the robot density of the automotive sector and all other industry sectors, for the countries with the largest robot populations. This demonstrates the significant potential for robots throughout worldwide industry. If the non-automotive industries applied robots in the same ratio as the automotive industry, this would result in an enormous growth in the number of installations. Even in the automotive industry, there is still significant potential for new robot applications, particularly in trim and final assembly operations, which have not yet been addressed. There is also very large potential within the developing economies, and particularly within China, Brazil and India. As the affluence and requirements of the local consumers grow, they will increasingly demand products that will drive the development of industries to provide those products, in turn leading to greater use of robotics.

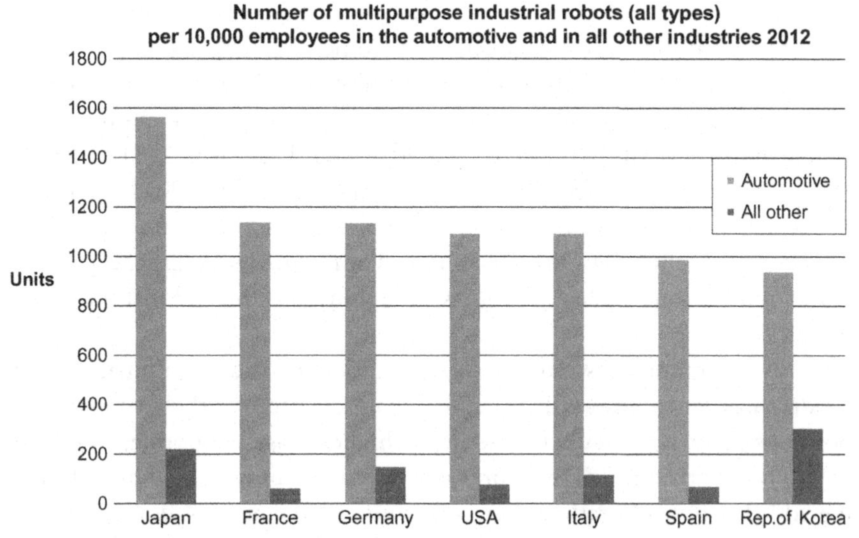

Figure 1.7 Robot density. *Source: IFR World Robotics, 2013.*

1.5 ROBOTS VERSUS EMPLOYMENT

One of the common misconceptions regarding robots is that incorporating them in the workplace leads to unemployment. It is often true that a specific robot installation will replace certain jobs. However, the automation of those tasks increases the competitiveness of the company and leads to growth, therefore producing more jobs. Normally, robots are employed for dirty, dangerous, or demanding tasks, the so-called 3Ds. Therefore, the jobs they replace are less suitable for workers. The new jobs that are created (e.g., programming or maintaining the robot systems) require higher skills, are more rewarding, and normally have higher rates of pay.

Ever since the Industrial Revolution, the trend regarding employment in industry has always been to reduce the number of workers directly employed in production. Every new development has required less direct manual input for the production process. The overall affluence created by successful manufacturing has supported the creation of jobs in service sectors. This trend will continue and robots are no different from other machines in this regard.

The improved quality and reduction in cost of manufacture leads to market growth because more people can afford and wish to buy the products. This again provides a positive effect on employment. As an example, if cars were still manufactured using the same techniques as 50 years ago, they would be less reliable and more expensive. Therefore, sales would be lower and the automotive industry smaller. For some products, such as the latest smart phones and tablets, the manufacture of these at prices consumers could afford would not be possible without the use of automation and robots.

All of these factors have been captured by an IFR study in 2011 concluding that the 1 million robots in use throughout the world in 2011 had led to the creation of nearly 3 million jobs. That is nearly three jobs for every robot. This did not include jobs that were created or retained in indirect roles such as the distribution or sale of products. The study also forecast that this trend would continue.

Thus, all companies must actively consider the use of automation and robots in order to ensure that they maintain or improve their competitive positions. This is particularly true in the higher-wage economies in which the skills and attributes of the workforce need to be fully utilised in performing added-value tasks rather than performing mundane jobs. Today's economy is truly global, and all businesses need to be aware that their competition, which may well be overseas, could be improving their operations and strengthening their competitive position. Ever-increasing health and safety requirements also demand that attention is paid to those tasks that are both arduous and dangerous.

The drive to apply robots will increase, and all companies need to develop the skills and expertise required to identify automation opportunities and to implement the appropriate solutions. The experience of those who have implemented successful robot systems is often based on lessons learnt from mistakes that have been made in the past. By the application of the correct approach, as outlined in this book, these mistakes can be avoided, and robot systems can be installed in ways that meet the expectations of the user. These then provide a positive contribution to manufacturing operations, improving both the product and the workplace, for the benefit of all the stakeholders in a business, including the workforce, the owners, and the customers of that business.

CHAPTER 2

Industrial Robots

Chapter Contents

Abstract

This chapter provides more detail on industrial robots commencing with the accepted definition. The various configurations are introduced, including articulated, SCARA, cartesian, and parallel or delta. The typical applications and market shares for each configuration are discussed. The key issues regarding robot performance, including working envelope and repeatability, are discussed together with the main points to consider when selecting robots. This includes a review of the typical contents of a robot data sheet. The benefits that robots can provide are also discussed, both for system integrators and end users. This includes the 10 key benefits that robots can provide for a manufacturing facility.

Keywords: Robot configuration, Articulated, SCARA, Cartesian, Parallel, Delta, Working envelope, Repeatability

An industrial robot has been defined by ISO 8373 (International Federation of Robotics, 2013) as:

> *An automatically controlled, reprogrammable, multipurpose manipulator programmable in three or more axes, which may be either fixed in place or mobile for use in industrial automation applications.*

Within this definition, further clarification of these terms is as follows:

- Reprogrammable – motions or auxiliary functions may be changed without physical alterations.
- Multipurpose – capable of adaptation to a different application possibly with physical alterations.
- Axis – an individual motion of one element of the robot structure, which could be either rotary or linear.

In addition to these general-purpose industrial robots, there are a number of dedicated industrial robots that fall outside this definition. These have been developed for applications such as machine tending and printed circuit board assembly and do not meet the definition because they are dedicated to a specific task and are therefore not multipurpose.

As mentioned in Chapter 1, the first application of an industrial robot was at General Motors in 1961. Since that time, robotic technology has developed at a fast pace and the robots in use today are very different to the first machines in terms of performance, capability, and cost. There have been various mechanical designs developed to meet the needs of specific applications, which are described below.

These different configurations have resulted from the ingenuity of the robot designers combined with advances in technology, which have enabled new approaches to machine design. The most significant of these was the introduction of electric drives to replace the use of hydraulics and the increasing performance of the electric drives, providing increased load-carrying capacity combined with high speed and precision.

Initially, hydraulics was used as the primary motive power. Hydraulic power was capable of providing the load-carrying capacity necessary for the early spot welding applications in the automotive industry. However, the responsiveness was poor and the repeatability and path following capabilities limited. For the first installations the robot technicians were required to start work early to turn on the robots, so they were warmed up prior to production starting, to ensure the robots performed repeatably from the first car body to welded.

Pneumatics were used to provide a low cost power source; however, this again could not achieve high repeatability due to the lack of control available. Hydraulics were also used for the early paint robots because electric drives could not, at that time, be used in the explosive atmosphere of the paint booth, caused by the use of solvent-based paints. Painting, by the nature of the application, carrying a spray gun with a 12 inches wide fan, about 12 inches from the surface, did not require the repeatability and control necessary for other applications; therefore, this proved to be a successful application for robots.

Electric drives of various different types have been used. DC servo motors were initially the most prevalent. These however had limited load-carrying capacity, which did initially provide constraints for the use of robots for spot welding applications due to the weight of the welding guns. Stepper motors were also utilised for high precision, low load-carrying applications. Once AC servo motors became available these took over the majority of applications. Their performance has continually increased providing better control, high repeatability, and precision as well as high load-carrying capacity. AC servo motors are now utilised in almost all robot designs.

2.1 ROBOT STRUCTURES

An industrial robot is typically some form of jointed structure of which there are various different configurations. The robot industry has defined classifications for the most common and these are:

- Articulated
- SCARA
- Cartesian
- Parallel (or Delta)
- Cylindrical.

These structures and their benefits are described in more detail below. The structures are achieved by the linking of a number of rotary and/or linear motions or joints. Each of the joints provides motion that collectively can position the robot structure, or robot arm, in a specific position. To provide the ability to position a tool, mounted on the end of the robot, at any place at any angle requires six joints, or six degrees of freedom, commonly known as six axes.

The working envelope is the volume the robot operates within. This is typically shown (see Figure 2.1) as the volume accessible by the centre of

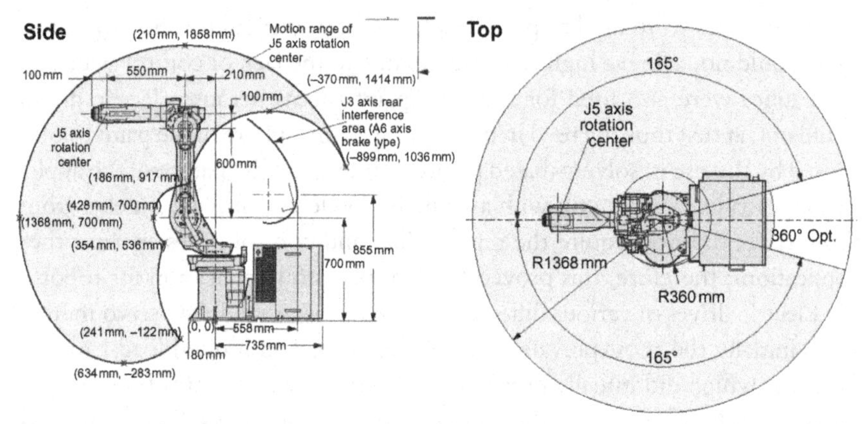

Figure 2.1 Typical working envelope.

the fifth axis. Therefore, anywhere within this working envelope the robot can position a tool at any angle. The working envelope is defined by the structure of the robot arm, the lengths of each element of the arm, and the motion type and range that can be achieved by each joint. The envelope is normally shown as a side view, providing a cross-section of the envelope, produced by the motion of axes 2–6 and a plan view then illustrating how this cross-section develops when the base axis, axis 1, is moved. It should also be noted that the mounting of any tools on the robot will also have an impact on the actual envelope accessible by the robot and tool combined.

The first robot, a Unimate, was designated as a polar-type machine. This design was particularly suited to the hydraulic drive used to power the robot. The robot (Figure 2.2) provided five axes of motion; that is, five joints that could be moved to position the tool carried by the robot in a particular position. These consisted of a base rotation, a rotation at the shoulder, a movement in and out via the arm, and two rotations at the wrist. The provision of only five axes provided limitations in terms of the robot's ability to orientate the tool. However, in the early days, the control technology was unable to meet the needs for six axes machines.

2.1.1 Articulated Arm

The most common configuration is the articulated or jointed arm (Figure 2.3). This closely resembles the human arm and is very flexible.

Figure 2.2 Unimate robot.

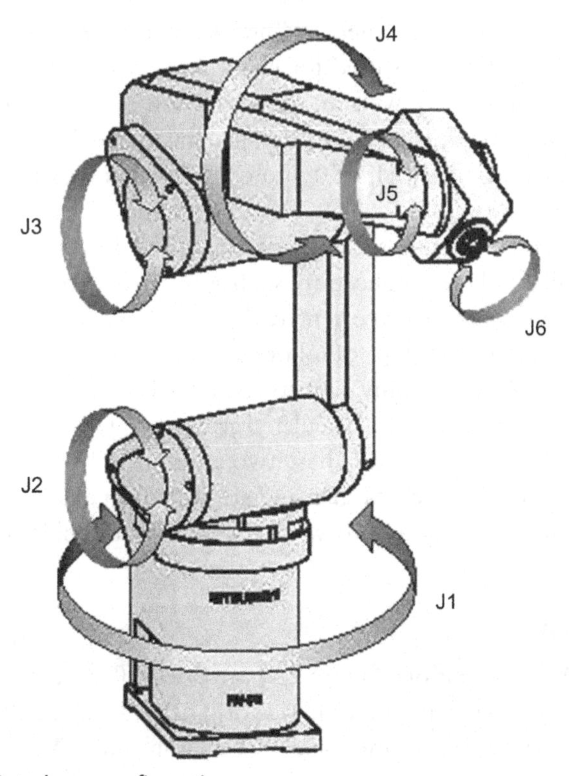

Figure 2.3 Jointed arm configuration.

These are normally six axis machines, although some seven axis machines are available, providing redundancy and therefore improving access into awkward spaces. The structure comprises six rotational joints, each mounted on the previous joint. They have the ability to reach a point, within the working envelope, in more than one configuration or position a tool in any orientation at a specific position.

The joint motion of articulated arm robots is complex and therefore can be difficult to visualise. The construction of the arm means each joint has to carry the weight of all the following joints; that is, joint three carries 4, 5, and 6. This impacts both the carrying capacity, the load that can be handled by the robot, as well as the repeatability and accuracy (see Section 2.2). The structures are not particularly rigid and the overall repeatability is the cumulative of all of the axes. However, the increasing performance of AC servo motors and the improvement in the mechanics provide excellent performance for the majority of applications.

As mentioned, the articulated arm is the most common industrial robot structure providing about 60% of annual installations worldwide, although it is higher in Europe and the Americas (International Federation of Robotics, 2013). This type of robot is used for many process applications, including welding and painting, as well as many handling applications including machine tool tending, metal casting, and general material handling. Typical robot sizes range from a reach of 0.5 to over 3.5 m and carrying capacities from 3 to over 1000 kg.

There are also a number of four axis articulated arms. These have been developed specifically for applications such as palletising, packing, and picking where it is not necessary to orientate the tool. Therefore two of the wrist axes are not required. This type of robot is able to achieve higher speeds with higher payloads than the equivalent six axes machines.

Dual arm robots, with two articulated arms mounted on the same structure, are also being developed. These two arms are able to work cooperatively and therefore mimic a human and are aimed at tasks such as assembly where two hands are required to work together to assemble the parts.

2.1.2 SCARA

The SCARA configuration (Figure 2.4) provides different attributes to the articulated arm. This configuration was originally developed for assembly applications, hence the name Selective Compliance Assembly Robot Arm. The four-axis arm includes a base rotation, a linear vertical motion

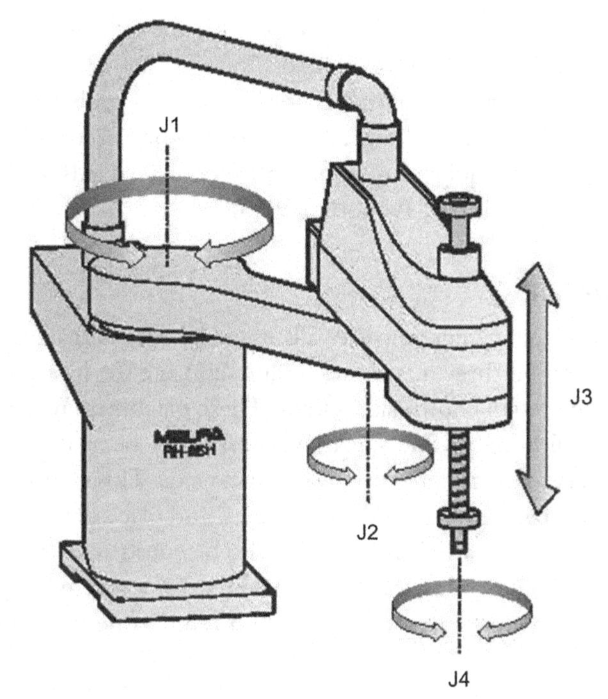

Figure 2.4 SCARA configuration.

followed by two rotary motions in the same vertical plane. Due to the nature of the configuration the arm is very rigid in the vertical direction and can also provide compliance in the horizontal plane. It provides high speed combined with high acceleration and works to very tight tolerances.

SCARA machines are typically small with the largest having carrying capacities of about 2 kg and a reach of about 1 m. They are mainly used for assembly applications although they can also be used for packing, small press tending, adhesive dispensing, and other applications. The application of SCARAs is mainly constrained by their size and the limitation of being only four axes.

A standard cycle time test for robots has been defined to provide comparability between machines. This so-called goalpost test consists of a 25 mm vertical move upward, followed by a 300 mm horizontal move, and a 25 mm vertical move down and simulates a typical move for an assembly application. The timings for SCARA robots to achieve this move, both forward and return, can be as low as 0.3 s. This is normally faster than the equivalent articulated arm, six-axis robots.

The SCARA configuration makes up about 12% of global sales although it is more popular in Asia, due to the size of the electronics sector in this region. Asia accounts for about 50% of all SCARA robot sales (International Federation of Robotics, 2013).

2.1.3 Cartesian

The cartesian category encompasses all industrial robots that include only linear drives for their three major axes (Figure 2.5) and the motions are coincident with a cartesian coordinate system. These machines are often limited to three axes, although some special versions have been developed with additional rotary axes mounted on the last linear axis. This cartesian category includes gantry machines as well as linear pick and place devices. The configuration of these are varied and they can also be constructed from modular kits, providing the flexibility to design a machine for a specific requirement. Gantries can be goalpost type devices, supported on one structure only, as

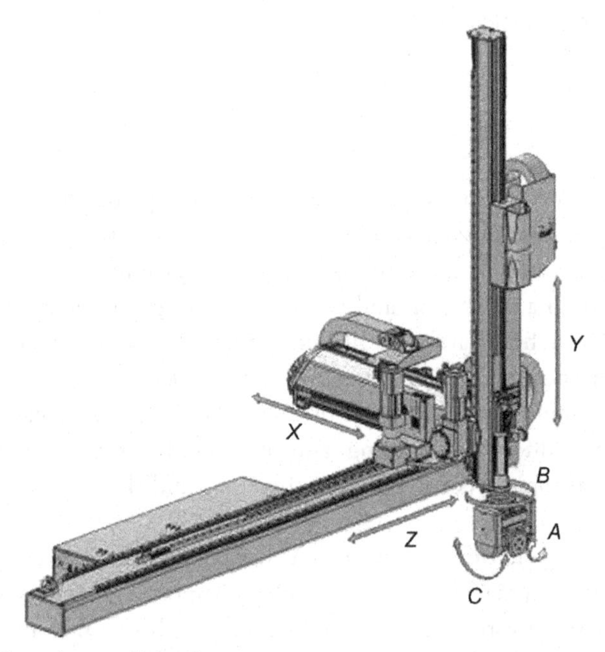

Figure 2.5 Cartesian configuration.

well as area gantries with two support structures. The main axis can range in length from less than 1 m to many tens of metres. Gantries can also be very heavy duty, able to carry 3000 kg. A further benefit of gantries is that they minimise the impact on the factory floor and manual access to machines, as they are largely overhead. However, they are often more expensive than the equivalent articulated arm robots.

Applications are quite varied although they are typically used for handling, palletising, plastic moulding, assembly and machine tending. They also have some application for processes such as welding and glueing, particularly on very large parts. Cartesian machines are the second most popular configuration, taking about 22% of global robot sales (International Federation of Robotics, 2013).

2.1.4 Parallel

The parallel or delta robot configuration (Figure 2.6) is one of the most recent configuration developments. This includes machines whose arms have concurrent prismatic or rotary joints. These were developed as overhead mounted machines with the motors contained in the base structure driving linked arms below. The benefit of this approach is that it reduces the weight within the arms and therefore provides very high acceleration

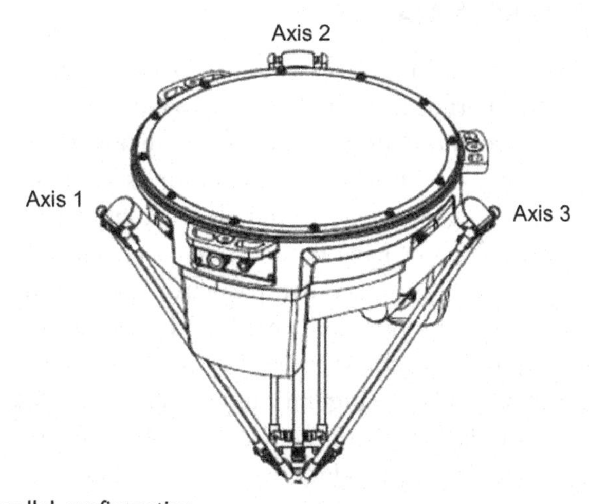

Figure 2.6 Parallel configuration.

and speed capability. However they do have a low payload capacity, typically under 8 kg.

Therefore, the main application is picking, particularly on packing lines for the food industry, and also assembly applications. These machines can achieve similar cycle times to the SCARAs with the fastest achieving the goalpost test (25, 300, 25 mm) in 0.3 s. This type of robot is sold in relatively small numbers, achieving only about 1% of the global market (International Federation of Robotics, 2013).

2.1.5 Cylindrical

These robots have a combination of rotary and linear axes, typically with a base rotation followed by a vertical and horizontal linear axis and further rotary axes at the wrist. They provide a rigid structure, with good access into cavities and are easy to programme and visualise. However, they do require clearance at the rear of the arm. They are particularly suited to machine tending and general pick and place applications.

They are mainly used in the electronics industry, particularly clean room applications, and take about 2% of the global market. Similar to the SCARA, they are most popular in Asia, due to the strength of the electronics sector in that region, which takes about 90% of global sales (International Federation of Robotics, 2013).

2.2 ROBOT PERFORMANCE

The type of structure and number of axes does have significant implications regarding the performance of a robot, which, as discussed above, does tend to make certain configurations more suitable for specific applications. For example, the SCARA configuration is particularly suited to assembly tasks requiring high speed and repeatability. The robots produced by each manufacturer have different characteristics and capabilities based on the applications for which that particular robot model is targeted. The main robot manufacturers produce a broad range of robots covering the capabilities required to address a range of applications. Originally they specialised in one particular configuration, articulated, SCARA, or cartesian structures and today they still tend to focus on that main configuration, although most do also produce robots with different structures to address the requirements of specific applications. There is often a wide choice of different robots from different manufacturers for any one application.

In addition to the number of axes and the configuration, the main performance characteristics of a robot are defined by the four parameters:

- Weight carrying capacity
- Repeatability
- Reach and working envelope
- Speed.

The weight-carrying capacity is normally the maximum load that can be carried at the tool-mounting flange of the robot wrist. With this load the robot will meet all other specifications, including repeatability and speed as well as providing reliable long-term operation. It should be noted that the detailed specifications, often provided in the robot manual, specify the position of the centre of gravity of this load in two directions from the tool-mounting flange (see Figure 2.7). As the distance from the tool-mounting flange increases the available load capacity reduces and, therefore, the ability of a robot to carry a tool or part, even if the weight is below the specified load capacity, may be compromised if the tool is relatively large.

The repeatability as specified for a robot is normally the point repeatability but in some cases path repeatability is also stated. It should be noted that robots, although repeatable, are not inherently accurate. Most robots, because of their structure, are not able to move to a commanded position accurately, for example, an *XYZ* coordinate in space, but will consistently repeat a taught position, within the tolerance band specified by their repeatability. The point repeatability is useful for spot welding, handling, assembly and similar types of applications but for process applications, such as welding and adhesive dispensing, the path repeatability is more useful.

The reach and working envelope is normally defined to the centre of the wrist axes. For a six-axis robot this is to the centre of axis 5, which means the robot can orientate the wrist to the full capability of its range even at the extreme of the reach or the edge of the defined working envelope. The working envelope is normally shown as a side and plan view (Figure 2.1). The robot should be able to reach any point within the working envelope. It should be noted that the shape of the working envelope will be different for different robot configurations.

The speed is often shown as the maximum speed achievable by each individual axis. This has limited value because the axes do not operate

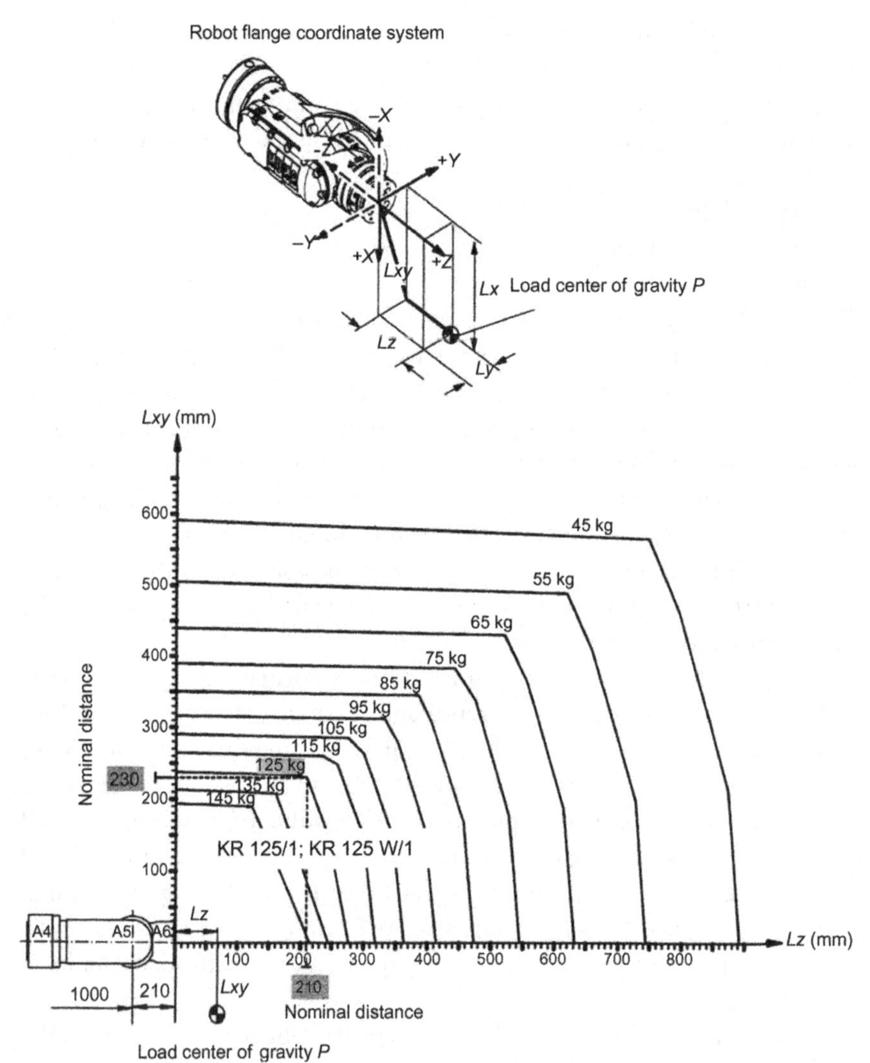

Figure 2.7 Robot load capacity.

independently and in many applications the moves to be undertaken are often short and do not allow the robot to achieve the maximum speed. However, the actual speed of the robot will influence the cycle time for many applications. The goalpost test (see Section 2.1.2) has been developed, specifically for assembly applications, to provide a reliable speed comparison between different robots.

A further point related to the speed and path repeatability is the tendency of robots to round off corners when moving at high speed. If the robot is programmed to move through a right-angle corner at lower speeds this can be achieved. If the speed is increased, the robot will round off the corner producing an error in the path. This error will increase as the speed increases. Solutions have been developed to address this problem, such as the ABB "TrueMove" control, which ensures the programmed path is achieved irrespective of the speed.

2.3 ROBOT SELECTION

The selection of a robot for a particular application is driven by the robot capabilities and performance required to meet the needs of the application as well as the requirements of the solution developed for that application. It should be remembered that there are normally a number of different ways by which the application can be addressed, for example a robot carrying the part to a fixed tool or a robot carrying the tool to a fixed part. Each of these approaches would require different capabilities in the robot. In this case, a different load-carrying capacity may well be required if the robot is carrying the part rather than the tool. The decision as to which route to take may well be driven by other aspects of the proposed system, such as the part handling in and out of the system or the overall cost of the alternative solutions. Therefore the process of robot selection is often iterative, requiring a number of different approaches to be considered before the optimum solution is defined. This is discussed further in Chapter 5.

Once the system concept is defined it is then possible to define the basic capability and performance characteristics required of the robot. These are typically defined by a data sheet, which provides the basic performance and capability information necessary to allow users to select the most appropriate robot for an application. Unfortunately, it is normally not a simple matter of working through the requirements of the application in comparison with the robot specification because a number of the parameters are linked. For example, the robot configuration defines the shape of the working envelope and the robot mounting position then defines how the working envelope is available for use which in turn defines the required reach.

Rarely is there only one parameter that stands out as the most important and therefore forms the starting point. As mentioned above, the normal starting point is a combination of reach, weight capacity, and repeatability. Ultimately the cost and simplicity of the overall solution is normally most important and therefore the robot selection and how it is to be applied should be aimed at achieving this objective (see Chapter 5).

A data sheet would normally cover the main specifications, including:

- Configuration

  The robot arm is normally shown as a photograph and/or a schematic that illustrates the type of structure, such as articulated, SCARA, or delta.
- Number of axes

  The number of axes is stated or can be determined from other data on the sheet, such as the working ranges and speeds.
- Reach

  This is normally stated as well as being illustrated on the working envelope. There may be variants of the model with different reach capabilities shown on the same data sheet.
- Working envelope

  Normally illustrated with a diagram showing both side and plan views.
- Payload

  The maximum payload on the wrist is normally stated. The data sheet may show variants of the model with different load capacities (different reach variants often have different load capacities). Normally the diagram specifying the payload at different offsets from the wrist (see Figure 2.7) is not included on the data sheet. There is often the capability to add additional load to the robot arm (axes 2 and 3) and this is sometimes, but not always, stated on the data sheet.
- Repeatability both position and path

  Point repeatability is normally stated.
- Axes working ranges and speeds

  The working range and maximum speed for each axis is normally stated.
- Robot mounting capabilities

  If the robot can be positioned in various orientations, such as wall and inverted mounting, this is stated.
- Dimensions and weight

  The weight of the robot arm is normally stated and the dimensions of the arm may indicated.
- Protection and environmental capabilities

  This would include the IP rating for the standard arm as well as any options. There may also be other options specified such as clean room (for electronics), washdown (for food), or foundry (for hot, dirty applications).

- Electrical requirements

  The supply requirements and power usage may be stated.

In addition to the characteristics of the robot arm, the capabilities of the robot controller will be important and these are often defined in a separate data sheet. However, for the initial development of concept solutions, the most important functionality relates to the robot arm. The importance of the controller relates more to the number of axes that can be controlled and the interfacing capability.

2.4 BENEFITS OF ROBOTS

The initial applications of robots were implemented by the early adopters who wanted to test the technology and determine what it might do for their businesses. Decisions to purchase were not necessarily made using stringent return on investment criteria but more on a belief that robot technology showed promise and the purchasers felt it important to be at the cutting edge of the application of that technology.

The increasing use of robots has been driven not by this interest in technology but by much more short-term financial criteria. Most companies today applying robots are basing their investment decisions on financial rewards that can be quantified prior to the investment and also realised once the system is in operation. It should be noted that some but not all of the benefits that can be realised by the use of robots can also be attained by the use of less flexible automation.

The potential benefits that can accrue from the application of robots fall into two categories: those that benefit the end user and those that benefit the automation solution provider. There has been much debate and investigation, largely driven by the robot suppliers, regarding the key benefits from the application of robots. These suppliers obviously have self-interest at heart, but the list of potential benefits has been refined and the International Federation of Robotics published a list of the key 10 benefits in 2005 (International Federation of Robotics, 2005).

These key benefits are as follows:

1. Reduce operating costs.
2. Improve product quality and consistency.
3. Improve quality of work for employees.
4. Increase production output rate.
5. Increase product manufacturing flexibility.
6. Reduce material waste and increase yield.

7. Comply with safety rules and improve workplace health and safety.
8. Reduce labour turnover and difficulty of recruiting workers.
9. Reduce capital costs.
10. Save space in high value manufacturing areas.

It should be stressed that these are potential benefits and may not be applicable in all cases. These are discussed in more detail below. First, the benefits for the automation system provider are discussed.

2.4.1 Benefits to System Integrators

On the presumption that the end user has already been persuaded of the benefits of automation and has decided to consider the purchase of a system, the automation provider or system integrator has the option of basing the solution on robots or other forms of automation. There is an increasing trend within the integrator community to utilise robots at the heart of their solutions. This trend is driven largely by two factors: the flexibility of the robot coupled with the fact that the robot is a standard product.

Robots are standard products and therefore offer measurable and known performance and reliability. As with all standard products, the designs undergo extensive testing prior to release and are also built to published standards. Therefore, a system integrator can review a product catalogue to review the basic specifications, such as load capacity and working envelope, and select a robot that meets the needs of the application (see Chapter 5). The major robot suppliers have an extensive range of robot types and, therefore, there are often a number of alternative machines that can be selected.

These published characteristics allow the system integrator to purchase a machine to meet the needs of its application for a known price. Although it is normally feasible to develop a bespoke machine to achieve the same characteristics, this is more of an unknown in terms of the development time and, therefore, cost. Similarly, the robot being a standard product provides known reliability performance, including mean time between failures (MTBF) and mean time to repair (MTTR), whereas a bespoke machine will always be an unknown quantity. Therefore, the longer-term risk to the end user and, as a consequence, the automation supplier is higher for the bespoke system.

The other major benefit of the robot is the flexibility it provides. The robot comprises multiple axes, often 6 axes, and can therefore orientate parts or tools as required. More importantly the motion of the robot is programmed through software. It is therefore easy to modify the robot operation even at a late stage in the project with limited cost implications. This flexibility can accommodate both late design changes in the product and also mistakes in the design and construction of the automation system.

Both of these factors – the flexibility and the known performance – reduce the risk for the automation solution provider. It is therefore possible to reduce the sales price, as any financial contingency can be reduced, which also increases the likelihood of an order from the end user. Once an order has been received, the timescale of the project may also be reduced, as the delivery of a standard robot may be quicker than the time required to design and build bespoke equipment.

2.4.2 Benefits to End Users

As mentioned above, there are 10 key benefits applicable to end users. In many cases, potential end users only consider the cost benefit of the direct labour saved as a result of the automation but often the financial justification (see Chapter 7) can be improved by also assessing all potential benefits and, in some cases, significant savings from another benefit may outweigh the saving in labour cost. These benefits and their applicability are discussed below.

Reduce Operating Costs

A robot system can help to reduce the operating costs associated with the manufacture of specific parts. This would include direct costs, such as the labour being replaced. However overhead can be also be reduced.

Energy saving can be achieved from a number of sources. The energy per unit of output is optimised by the consistency of the robot and the resulting reduction in scrap or rework. Additionally, robots do not require the same heating (or cooling) and lighting necessary for manual labour. Therefore, if the particular area of the factory can be fully automated it is then possible to reduce energy consumed to maintain the working environment.

The reduction in direct labour may also have an impact on other aspects of the operation, such as training, health and safety, and employee administration. Less indirect labour is required to sustain and manage the smaller workforce and, therefore, savings in overhead costs can be realised.

Improve Product Quality and Consistency

Robots are repeatable. They will consistently produce output that will be of high quality if they are correctly set up and fed consistent input materials. They do not suffer from the inconsistencies of manual labour caused by tiredness, boredom, or distraction that is often associated with tedious or repetitive tasks. This inconsistency results in variable output and variable quality. In contrast, the robot will provide regular output; that is, a known number of manufactured parts per day as well as consistent quality, meaning a known number of good parts per day.

Improve Quality of Work for Employees

Robots can help improve the working conditions for the employees. They can take over dirty, dangerous, and demanding tasks, the so-called 3Ds. Examples of these is paint spraying, working on presses, or handling heavy loads. These are often tasks where it is difficult to maintain consistency, so the application of robots will also improve the production output or the quality of parts produced. The use of robots allows manual labour to be removed from direct involvement in the production process, thereby reducing the pressure on the workers to maintain a regular output.

The operation and maintenance of the robots also requires enhanced skills and, therefore, the motivation of the staff can be improved by enhancing their skills and increasing their pay at the same time as improving their job.

Increase Production Output Rate

As mentioned above, the consistency of robots ensures a regular production output. It also means that the output of other machines can be maximised as the robots will be always be ready to perform their function as and when necessary. For example, machine tools can be unloaded and reloaded exactly when ready rather than waiting for an operator who may be distracted, on a break or involved in some other activity.

It is also possible to run some operations on extended shifts or overnight and on weekends without manual intervention. Again, expensive machine tools can be set up to run overnight using robots to unload and reload, providing additional output for limited additional cost. This enhanced production capability can be particularly valuable for subcontractors who may have variable orders from their customers.

Increase Product Manufacturing Flexibility

Robots are inherently flexible, more so than other forms of automation. Once an operation is programmed into the robot, it can be recalled and be operational within seconds. Therefore changeovers can be very quick, minimising downtime. A robot system can handle variants of the same product or even different products, providing the opportunity for small batch sizes.

There can be constraints caused by the equipment around the robot, such as fixtures or grippers, but by careful design and good concepts these can be overcome. However, it should be recognised that the robot is not as flexible as a human operator.

Reduce Material Waste and Increase Yield

One of the major benefits of robots is their consistency. By ensuring quality output, more "right first time" products are produced and, therefore, material waste is reduced. In addition, the robots will use less consumables to produce the parts; for example, a robot welder will always produce the size of weld required, not too big and not too small. A robot painting system will always apply the same thickness of paint.

The introduction of automation, such as robots, also imposes consistency on the parts or products being fed to the automation, providing an overall benefit to the business. The automation can also be used to check the incoming product and, therefore, identify when product is out of specification. This can be particularly valuable in the food industry where output product must meet minimum weight requirements. Robots and automation can be used to drive down the tolerance band used to ensure the so-called give away, that is the free product provided to the consumer, can be minimised.

Comply with Safety Rules and Improve Workplace Health and Safety

Robots can take over unpleasant, arduous, or health-threatening tasks currently handled by manual workers. There is increasingly stringent health and safety legislation that makes some of these tasks either difficult or impossible to perform manually and robots provide an increasingly cost-effective alternative.

By removing operators from direct contact with machines or potentially hazardous production machinery or processes, robots can decrease the likelihood of accidents. For example, the processing of forgings through foundry presses is both an arduous and dangerous operation that can in most cases be automated by the use of robots. There are many repetitive or intensive industrial activities that can lead to ailments, such as repetitive strain injuries (RSIs) and vibration white finger; for example, the polishing of metal parts or the fettling of castings. The use of robots can remove the operators from these tasks, reducing the risk of injury.

Reduce Labour Turnover and Difficulty of Recruiting Workers

Improving the working environment and also removing the most repetitive or demanding jobs will reduce the labour turnover within a business. The workforce is more likely to be satisfied if it is given more challenging, less repetitive roles that also require higher levels of skill, particularly if these enhanced roles also lead to higher levels of pay.

If staff retention is improved, the cost associated with hiring new workers is reduced. This would include not only the costs directly associated with the hiring process but also training costs and the cost of lower productivity as workers come up to speed on the job.

Reduce Capital Costs

The flexibility provided by robots can reduce capital costs by enabling small batch sizes, reducing the cost of work in progress and inventory. Robots can provide the opportunity to run other machinery more efficiently or perhaps over extended hours, which may mean it is unnecessary to purchase additional machines. For example, one machine tool operated by a robot overnight on a lights-out basis could remove the need to purchase an additional machine tool.

Robot systems, because of their flexibility, are also reconfigurable and can be reused if products or product designs change, whereas more dedicated automation may need to be replaced.

Save Space in High-Value Manufacturing Areas

Robot systems can be very compact as they do not require the same space as an operator performing an equivalent task. Robots can also be mounted in various attitudes, such as wall or overhead, to reduce the space required. This provides the opportunity to minimise the floor space required, which in some cases can be a valuable benefit.

By careful assessment of each of the above benefits, it is often possible to build a case for the application of a robot system with a significantly improved financial return than can be achieved by considering only the labour costs. Further discussion relating to return on investment and the justification of automation is provided in Chapter 7.

CHAPTER 3

Automation System Components

Chapter Contents

Abstract

An automation system will include different components, in addition to the robot, to provide the complete solution. This chapter provides an introduction to some of the most frequently used technologies, including handling and feeding systems, vision, grippers and tool changers, as well as tooling and fixtures. The process equipment required for a robot application is discussed, particularly for welding, painting, dispensing, and material removal applications. Assembly automation is also reviewed. The basic requirements for the system control, including networks and human machine interfaces, are introduced together with the basic principles of safety and guarding for automation systems.

Keywords: Bowl feeders, Machine vision, Dress package, Positioner, Grippers, Tool changers, Fixtures, Assembly automation, PLC, Network, Safety

Each automation system is configured to meet the specific requirements of the application for which it is intended. As will be seen in Chapter 4, there are many different types of automation systems for a wide range of applications across almost all manufacturing sectors. The application requirements for automation in the food industry are very different to the requirements for the electronics sector. Therefore, there can be many different automation components that need to be brought together to successfully address the requirements for a specific application. These components also vary significantly between each application; for example, a gripper for picking soft fruit is very different from a gripper for handling hot forgings.

It is not possible to provide an exhaustive appraisal of all the various types of automation components due to the wide range and varied specifications and capabilities this would cover. However, the main components can be reviewed and the most important features of each of these discussed. The intention is to provide an overview of the more common elements of automation systems and also the most important issues that should be considered when using these components.

This chapter covers the following:

- Handling equipment
- Vision systems
- Process equipment
- Grippers and tool changers
- Tooling and fixturing
- Assembly automation components
- System controls
- Safety and guarding.

The development of an automation system, by the integration of these automation components together with robots, is discussed further in Chapters 4 and 5.

3.1 HANDLING EQUIPMENT

In any manufacturing facility it is critical that materials be delivered to the various operations as required for the efficient processing of the parts through that facility. The objectives for any material handling operation are getting the right parts to the right place, at the right time, in the right quantity, and also avoiding damage to these parts. The consequences of a poorly designed or inappropriate material handling system include: excess work in progress, disorganised storage and poor inventory control, excessive handling, product damage or excessive scrap and idle machines. It should always be noted that material handling does not add value to the product, it only adds to the cost. Therefore, the material handling needs to be as efficient as possible to minimise this cost.

There are many different types of equipment used, including pallet trucks, forklift trucks, overhead cranes, and conveyors, to provide for mass movement of parts around a facility. These are not considered in detail here. The objective of this section is to review those handling systems that can have a direct impact on the design and operation of an automation system. With any automation system, the feeding of parts to the system and the removal of completed parts can, if the handling is not appropriate, seriously affect the performance of the automation system. Therefore, the handling systems must always be considered when automation solutions are being developed to ensure the planned performance is not constrained by the input or output of the parts. Automation will not solve material handling issues. Conversely, automation will often highlight any material handling problems because the automation system will not perform as intended.

3.1.1 Conveyors

Conveyor systems can be used to move parts between automation systems and also within a single automation system; for example, the feeding of boxes from packing lines to robotic palletising systems or the transfer of parts within an assembly system. Conveyors provide for fixed movement between two points along a predetermined path. They can be on the floor, located above the floor or positioned overhead. The selection depends on the product to be moved, the space available, and the access required to the other equipment and operations. They are particularly suited to moving high volumes of product and can also provide temporary storage or buffers between specific operations.

The motion of conveyors is normally either continuous or indexing. Continuous conveyors do not stop, and the processes need to be performed on the moving line or the products need to be removed for specific operations to take place. Indexing conveyors are often used within assembly-type operations where the product will be stopped for each operation to be performed; however, these provide a constraint as the conveyor can only move when the slowest operation has been completed. An automated system may include a small number of conveyors; for example, to provide for the automatic removal of full pallets from a palletising system, or a number of different conveyors of different types and speeds to provide for the flow of parts or products through a complete system. Often, the conveying solution is determined by other operations. For example, a continuous conveyor is very efficient for a painting operation because the product can be reliably fed through the oven, via a continuous conveyor, to achieve a repeatable and efficient baking cycle.

There are many different types of conveyors including belt, chain, and roller conveyors. Belt conveyors are particularly suited to the movement of low weight and delicate items, such as food products. They are also more suited to applications where cleanliness is important, such as the food sector, but are limited to straight lines. Chain conveyors either carry the product direct on the chain or via carriers that are linked to the chain. These are suited to heavier products and slower speed applications, such as the transfer of pallets or the conveying of parts on fixtures through a paint shop. Roller conveyors are particularly suited to midrange parts in terms of size and weight, such as cartons, and are often used to feed packed product to palletising or storage systems.

3.1.2 Discrete Vehicles

Automated guided vehicles (AGVs) can achieve some of the benefits of conveyors but provide greater flexibility and less obstruction on the factory floor. However, they cannot handle the same throughput as a conveyor, do not inherently provide a buffer and are more expensive. An AGV is effectively an unmanned vehicle that is guided automatically around the factory floor. This can be achieved by a buried wire, which the vehicle can sense and follow.

It is feasible to provide different routes and junctions that can be selected by the overall control system to allow the AGV to deliver products via different routes to multiple operations. The AGVs often operate in the same working area as other users, such as forklift trucks and the workforce; therefore, they include appropriate safety systems. AGVs are, however, relatively slow and are limited in terms of the size and weight of product that can be handled. Because they are guided by wires buried in the floor it can be expensive to change or expand an AGV system. Self-guided vehicles are also available that use wall-mounted targets and laser scanners or an internal GPS system to determine their location. These self-guided vehicle systems are less constrained, simpler to modify and, therefore, provide greater long-term flexibility.

3.1.3 Part Feeding Equipment

In all automation systems, there is a need to feed individual parts into the system. The most effective approach may be the manual loading of parts into fixtures (see Section 3.5); for example, the loading of sheet metal components into a fixture for a robotic welding system. Alternatively, the feed may be via a conveyor system bringing the parts from a previous operation. For some applications, particularly assembly, there is the need to feed individual components, in high volumes with high frequency, into the automation system.

The components being fed could either be parts to be assembled or the joining tool, such as screws or rivets. These feed systems must be highly reliable because the effectiveness of the automation system is totally dependent on the performance of the feed systems. The feeding system, in addition to providing the components to specific positions for subsequent operations, also needs to ensure the parts are orientated correctly for these operations. There are a number of techniques, including magazine and bandolier feeders (see below), which can be used but these require preprocessing to pack the components and provide the orientation prior to loading to the feeders. However, many components are delivered loosely in boxes or crates. The cost of the parts may be lower because no preprocessing is required but the automation system needs a mechanism to accept these parts in bulk and in random orientations, which can then sort the components and deliver them to the required positions in the desired orientation. The vibratory bowl feeder is the most common equipment used for this purpose with almost 80% of feeding to automated assembly systems being achieved using bowl feeders.

The design of the parts can have a significant impact on the ease with which they can be fed, which in turn drives the cost of the feeding system. If parts can be symmetrical, it minimises the sorting required. If some features are nonsymmetric, then it helps if this applies to the major features so, for example, the centre of gravity can be used to sort the parts. If open hooks are included there is a risk of tangling, whereas closed hooks will ensure tangling is not possible. It is therefore important that part design also considers the methods by which the parts are to be fed and processed by an automation system.

Bowl Feeders

Vibratory bowl feeders (Figure 3.1) consist of a bowl that is vibrated via an electromagnet. Within the bowl, a spiral track rises from the centre of the bowl around the inside of the periphery to the top. The vibration causes components within the bowl to rise up the track and, therefore, be fed out of the bowl. Towards the top of the track a selector mechanism, consisting of various mechanical features such as pressure breaks, wiper blades, and slots, is used to sort the components to ensure only those in the correct orientation are fed out of the bowl. Those that are incorrectly oriented drop back into the bowl to be recycled.

The size of the bowl is mainly dependent on the size of the components to be processed. Bowls can either be replenished manually or can be fed via external elevators taking product from a bulk hopper and depositing it into the bowl. The indexing of the elevator, providing a new quantity of

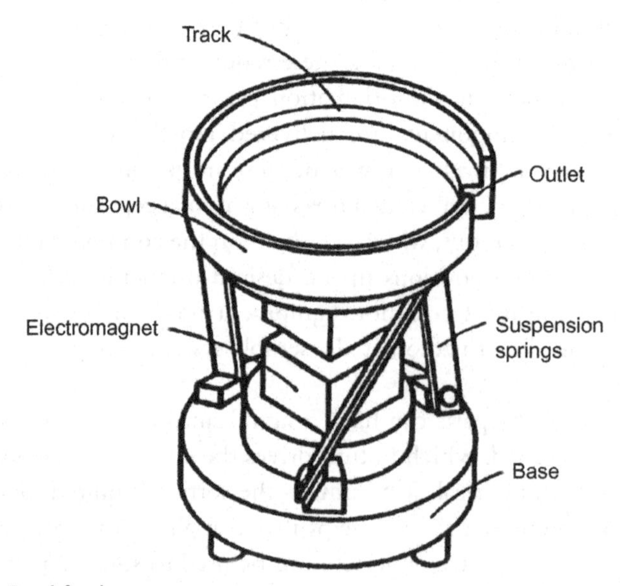

Figure 3.1 Bowl feeder.

product, is controlled to ensure the bowl always contains an appropriate number of parts.

Bowl feeders are simple devices and are highly reliable. They are by far the most frequently applied feed system and do provide sorting of parts from a random feed. Parts can be literally tipped into the bowl, and can handle a wide range of items, generally from small to medium size. The initial design of the bowl is critical both for the correct operation and the sorting. The gradient of the spiral track needs to be correct to ensure the parts rise up the track correctly, and the design of the selector mechanism needs to provide the required sorting. The design is based on the inherent characteristics of the components, including geometry and centre of gravity. The design is more an art than a science and is largely based on experience. The two main limitations of bowl feeders are they cannot handle parts that can be damaged by contact with other parts and also parts that can tangle, such as springs.

Linear Feeders

Linear feeders work in a similar way in that vibration provides the motion; although in this case in a straight line rather than the spiral of the bowl. These are often used in conjunction with a bowl feeder, taking the parts output from the bowl and feeding them to the required position within the automation system. Linear feeders can also feed larger parts, or more delicate parts, than can be fed via a bowl feeder; however, they do not provide any sorting, so the parts must be oriented correctly by another device, normally before the linear feeder. Alternatively, an intelligent picking system, possibly using vision (see Section 3.2), can determine the part position and orientation to provide for picking from the feeder.

Blow Feeders

Basically the part is blown down a tube that provides for fast delivery of parts but is limited to small parts. The use of the tube also provides the opportunity to feed components over a relatively complex or variable path; for example, to the end of a moving multiaxis device, such as a robot. This technique is typically used to feed screws or rivets to the application device, which could be mounted on a multiaxis mechanism to position it at locations on the component where the fixing is required.

Bandoleer Feeders

The components are held in bandoliers or tapes, which are then loaded to the automated machines. This technique is mostly used by the electronics industry to feed components to printed circuit board assembly machines. The bandoliers have to be premade, but the components are usually produced on automated systems that can relatively easily provide their output on bandoliers.

Magazine Feeders

Magazine feeders require parts to be prepacked either into trays or dispensers. Trays are suitable for parts that can be damaged by other feeding techniques or parts where the correct orientation is difficult to obtain. The trays can be moulded and are therefore low cost. It is often possible to use these trays to provide transport between manufacturing operations, normally stacked on pallets, providing for ease of transport and protection for the parts. The unloading of the tray into the automation system does require some form of handling device, which may be a robot or simple manipulator, to pick each part, or groups of parts, from the tray and place the part or parts into the required positions within the automation system.

Dispensers are often used in packing systems. These may be to feed card to case erectors producing boxes for the packing system or plastic trays, which are then being filled with the product. This might be the plastic packaging used for many food products. Dispensers are normally simple devices that hold a stack with a simple arm removing single items, picking using a vacuum gripper. These items are normally placed on a feed conveyor to provide transport into the packing automation.

3.2 VISION SYSTEMS

Machine vision is basically the use of optical, noncontact sensors to automatically receive and interpret an image of a real scene to obtain information and/or control machines or processes. Vision systems can be used in isolation, for example, as an inspection tool, or within an automation system. Initially vision systems, like most automation equipment, were expensive and difficult to implement and use. In recent years the cost has reduced significantly, vision capability has increased enormously, and they are now much easier to use. Therefore, the use of vision has grown exponentially and vision is now a widely used tool within many automated systems and processes.

It should be noted that machine vision in many respects does not yet match the capability of human vision and, therefore, careful consideration must be given to any vision application. Machine vision is consistent and tireless, may operate outside the visible light spectrum, can work in hostile environments, and precisely follows a predefined programme. Human vision, by contrast, has a much higher image resolution, can interpret complex scenes very quickly, is highly adaptable but is constrained to the visible light spectrum, does tire, and can be subjective.

Machine vision is suited for part identification, finding positions, inspection, and measurement. It is therefore used for applications such as inspection on high-speed production lines, microscopic inspection, closed loop process control in all environments including clean rooms and hazardous environments, as well as precise noncontact measurement and robot guidance. It is not intended to cover all these applications here but to focus on issues related to the use of vision within robot systems. The main applications of vision within robot systems are guidance, both for part picking and tracking, part presence/absence checking, defect identification and part identification, including optical character recognition and barcode reading. These will be discussed in more detail later.

First, it is worthwhile to briefly identify the main elements of a vision system and how it operates. A typical vision system would include the camera, lighting, processing hardware and software. The software is configured to tailor the vision system and the analysis performed to the specific application. There are three main operations within a vision system. The first is to obtain the image; second, to process or modify the image data; and finally to extract the desired information. Each of these operations has an impact on that which follows; for example, the method of lighting in the first operation can greatly simplify capturing the image, which then reduces the processing required and makes it easier to extract the desired information.

There is a wide range of cameras available, with the key parameters being resolution, field of view, depth of field, and focal length. The focal length determines the nominal distance at which the camera will provide a focused image, and the depth of field describes the range over which this focus is maintained. The field of view determines the size of the image taken at the focal length, and the resolution is the number of individual steps the image is divided into, which then determines the smallest measurement or feature that can be identified.

Lighting is most important. There are a number of different techniques available, including both direct and diffused lighting from the front, back, or side of the object as well as structured and polarised lighting. The effect of ambient lighting including: sunlight, general factory lighting, and light from any other sources must be considered. In particular, the impact of changes in ambient light must not affect the operation of the vision system. The purpose of the vision system lighting is twofold: first, to highlight the important features of the object and second, to remove any potential impact from changes in ambient light.

As an example, for weld guidance systems the vision sensor is mounted directly in front of the welding torch, looking at the weld seam only about 25 mm in front of the weld. To allow the camera to "see" the weld seam, the illumination is provided by a infrared laser-generated line and a filter is mounted in front of the camera to remove all the light with the exception of the wavelength of the laser. The light from the welding process is therefore removed from the image received by the camera, which then allows the camera to "see" the weld seam.

Back lighting can be very helpful for part location and measurement because it reduces the image of the object to a shadow, removing any features on the surface and, therefore, simplifying the task for the vision system. The background on which the object sits can also be important to help differentiate the part. A typical application for vision is to provide part location and orientation information when robotically picking parts from conveyors; for example, the packing of chocolates into boxes. White conveyors are often used because these provide a strong contrast between the colour of the chocolates and the conveyor.

The processing complexity and time required can be significantly reduced by highlighting the important features or removing unnecessary information from the image. Additionally, the reliability of the vision operation is also improved. The reliability is also improved if the effect of changes in ambient light can be removed. To fully remove the effect of ambient light it may be necessary to enclose the vision operation within a light-tight box.

Within robot automation systems, one of the most popular uses of vision is within packing applications, particularly within the food industry. Products are normally randomly positioned on conveyors being fed into a robot packing cell. The vision system is used to identify the positions of the products and feed the information to the robots to allow them to pick the product from the conveyor and place it into the packaging. These are normally moving conveyors and, therefore, the positions as measured by the vision system at the input need to be tracked through the cell to the picking point. These systems normally include multiple robots, so there is also an assessment made of which robot should perform the picking operation to balance the workload between the robots. There are standard solutions available for these types of applications, which have made their implementation much simpler and cost-effective.

The same vision system can also provide quality control; for example, by checking the shapes of the chocolates to be packed to ensure any misshapes are discarded. A further example is the packing of small pancakes where the vision also checks the colour of the pancake. Too dark indicates the pancake is overbaked or too light that it is underbaked; in both cases, the pancake is discarded. Vision is also used, particularly within assembly systems for the checking of features or parts within an assembly. This provides a check that a prior operation has been performed successfully and ensures an assembly does not continue with incorrect parts.

Vision has also been used to check that a fixture manually loaded with a number of different parts has all the parts loaded prior to the next operation to ensure everything is in place as required. This could be achieved with individual sensors for each part but the vision approach may be more cost-effective, particularly if there are a number of different variants processed through the same fixture.

Vision can be used to read characters on labels or barcodes to provide for product identification. For example, a palletising system may be receiving a mix of boxes with the vision system identifying the box to ensure it is placed on the correct pallet. In most applications of this type, a barcode reader would normally cost less but there are situations where a vision system is the more appropriate solution.

Machine vision by providing guidance, measurement, or quality control therefore enables the automation of applications that would not be otherwise feasible. The cost of vision systems continues to decrease and their ease of use and performance continues to improve. However, they do require careful investigation to ensure reliable operation.

3.3 PROCESS EQUIPMENT

Automation systems can be divided into two categories: those which handle parts for assembly, machine tending, or general material handling and those which apply some form of process. The former category tends to use grippers mounted on the robot, whereas the latter require some form of process equipment to control and apply the process. These process applications include:

- Welding.
- Painting.
- Dispensing of adhesives or sealants.
- Cutting and material removal.

Most of these processes can be accomplished manually, and the initial robot applications basically utilised the manual equipment mounted to the robot arm to replicate the manual application. However, over time the process equipment has been developed to suit the automated applications and techniques, and equipment has been developed that is more suited to the automated application. This has resulted in systems that are more closely integrated with robots, both mechanically and within the controls, to provide more effective, robust, and reliable solutions. It is not intended to provide a complete study of all the processes but to cover the most important issues related to the application of robots to these processes, as discussed below.

3.3.1 Welding

The main types of welding processes are spot welding and arc welding. Arc welding covers a number of process variants such as metal inert gas (MIG) and tungsten inert gas (TIG).

Spot Welding

Spot welding joins two sheets of metal together by applying a force to close the gap between the sheets and then passing an electrical current through the contact point causing fusion of the metal at the contact point. It is used extensively in the production of car bodies. The spot weld gun consists of two arms, each of which carry contact tips. The arms are driven, typically via pneumatics, to close the contact tips on the metal sheet providing the force to close the gap between the sheets. A transformer provides the electrical current, which passes through the circuit made by the arms and contact tips, generating the heat at the point of contact and thereby causing fusion. The key parameters controlling the process are the current supplied and the length of time it is applied.

The size and design of the arms of the welding gun is determined by the part to be welded. The contact tips need to be placed on each side of the flange to be welded and, therefore, the guns can often be large and unwieldy. This makes them difficult to orientate to the correct position manually; whereas robots can handle the weight and achieve the positioning required both quickly and accurately. As a result, spot welding became and remains a major robot application.

As the weight capacity of robots increased it became possible to integrate the transformer within the welding gun, which reduced the size of the cables to the weld gun. The welding controller, or weld timer, is also now integrated within the robot controller to provide a fully integrated package. The actuation to close the gun arms has also been developed and can now be achieved by a servomotor integrated within the weld gun. This is controlled within the robot programme in the same way as the other robot axes. The gun can, therefore, be more intimately controlled, not just open or closed, as with pneumatics, but to specific positions. This can reduce the time required, as the gun opening can be minimised and the closing can commence before the robot actually reaches the weld position.

The other important element of the spot welding robot is the dress package. This is the cable bundle that carries the services from the base of the robot to the weld gun. The cable bundle is the element that suffers the greatest wear because it is subjected to significant movement as the robot orientates the weld gun around the part. It is often the dress package that is the cause of downtime, rather than the robot or weld gun. Spot welding robots have been developed to integrate the dress pack, as far as possible, into the robot arm, including taking cables through the centre of some of the axes. There are also specific dress packages designed to suit spot welding robots to make the most of these features (Figure 3.2), the overall objective being to reduce the wear on the cables and increase the reliability of the complete spot welding system. It is also important that the weld tips be dressed on a regular basis, as they become deformed over time. Automatic tip dressers are available to provide this as a fully automated function, thereby minimising maintenance operations.

Arc Welding

Arc welding fuses two metal surfaces together using heat generated by an electric arc. Unlike spot welding, it only requires access from one side. The arc for MIG welding is produced using a wire, which also melts to add additional metal to the joint. TIG does not provide any additional metal,

Figure 3.2 Spot weld dress pack.

unless an external filler wire is used, and, therefore, the joint is formed purely by the fusing of the metal within the joint. Oxidation of the metal is prevented by the use of an inert gas to form a shroud around the arc. MIG welding is extensively used within automotive components, off-road vehicles, and general metal fabrications. TIG tends to be used for higher precision applications, particularly on thin metal. The majority of robot applications use the MIG process. The most important parameters are the feed rate of the wire, the voltage, and current applied as well as the stick out, the distance between the weld torch and the seam to be welded, and the speed of the weld torch along the seam.

A MIG welding robot will have the welding torch mounted at the end of the wrist. For many applications the torch is water cooled. The torch is fed with welding wire normally driven to the weld torch using a wire feed motor mounted on the shoulder of the robot. The wire supply is either a spool mounted on the side of the robot base or, more normally, a bulk pack located outside the robot cell. The dress package will also feed the inert gas and water to the welding torch. The design and mounting of the dress package is important to ensure reliability of the complete welding system as well as a reliable wire feed to the weld torch. Similarly to spot welding robots, robot arms and dress packages have been developed to minimise the wear on the cables, including the feed of the cables through the wrist directly into the weld torch.

The welding process is controlled via a welding power source normally positioned close to the robot. The control of the power source is fully integrated within the robot controller to allow the selection of the most appropriate welding parameters to suit the type of seam, the stick out, and the speed. These are normally programmed into tables to allow the correct parameters to be called at appropriate points within the robot programme. For larger joints, the robot can be programmed to perform weave patterns, using standard preprogrammed routines, to oscillate the weld torch across the seam as it welds, which assists the weld process.

Repeatable part fit up and position are key to successful welding. For larger parts, such as off-road vehicles, bridges, and other heavy fabrications, this is difficult to achieve. Techniques have been developed to provide robots with the capability to accommodate variations in parts. The simplest is touch sensing where the robot will use the tip of the welding torch to touch features close to the joint. The position is sensed as the torch is grounded and, thereby, determines the position of the joint. A second function is through-the-arc seam tracking. When the robot is weaving, across the joint, the weld current will vary as the stick out changes. By monitoring the current, the weld torch position can be maintained centrally along the weld joint. The use of this function is limited to certain types of joint and metal thicknesses, partly due to the need to weave. Therefore, vision systems have also been developed to provide tracking of the seam. However, these significantly add to the cost and complexity of a robot welding system.

It is often necessary to reorientate the part during the process to provide access for the robot to all the welds. For example, a car exhaust system consists of a number of pipes to be joined to a number of boxes. A full 360° weld around the joint between the pipes and the boxes is required. To achieve this, the parts need to be rotated to provide access for the robot. This can be achieved with a single axis servo-driven positioner. To achieve the best results, the motion of the 6-axis robot and the servo positioner need to be fully coordinated.

There are ranges of positioners with the simplest being a head and tailstock with a servo drive (Figure 3.3). These can then be built into two station positioners with two head and tailstocks being mounted onto a turntable (Figure 3.4). This provides the opportunity for the operator to be unloading and reloading one side whilst the robot is working at the other. Also, two axis positioners (Figure 3.5) are available where it is necessary to be able to orientate the part in two axes to provide access to achieve the optimum

Figure 3.3 Single axis positioner.

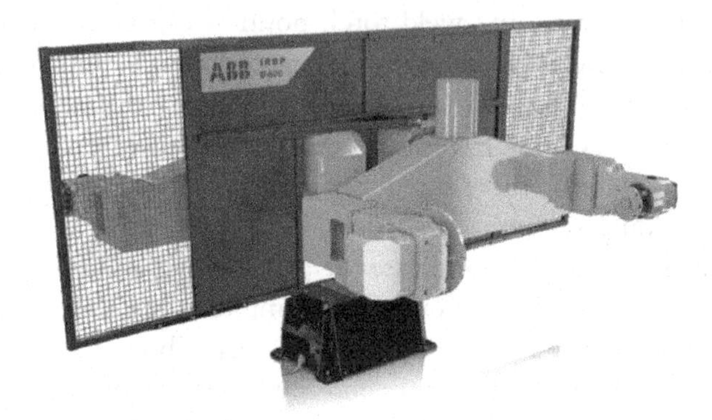

Figure 3.4 Two station positioner.

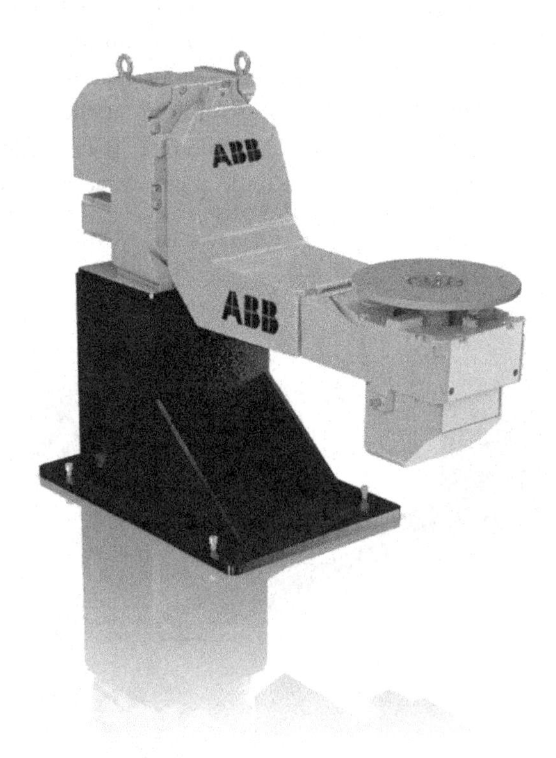

Figure 3.5 Two axis positioner.

weld positions. Again, these can be configured as two station positioners. Servo-driven robot tracks are also available if required for larger parts. Robots can also be mounted overhead in an inverted position, which in some cases provides improved access to the weld joints. There are many different combinations of positions and robots available, and the selection of these for typical applications is discussed in Section 5.2.1.

The final element within the robot welding package is the torch service station (Figure 3.6). During the welding process, the weld torch will become contaminated with weld spatter and will require cleaning on a regular basis to maintain the reliability and quality of the welding process. This is achieved by an automated cleaner consisting of a router to clean the inside of the weld

Figure 3.6 Weld torch service centre.

torch shroud and an oil spray to reduce the adhesion of the spatter. These torch service stations can also include sensing devices that check the position of the weld torch and correct for any misalignment that may have occurred due to collisions, wire stick problems (where the weld wire remains attached to the weld pool at the finish of a weld), or manual maintenance.

3.3.2 Painting

Painting tends to be considered as different from most other automated processes. This is in part from a historical perspective, in that the developers of automated paint solutions were in different businesses than those addressing other applications, but also because the requirements of the process and the language used are different from many other applications.

There are a number of different forms of automated paint equipment, such as reciprocators and electrostatic bell systems, but within the confines of this book we will concentrate on robotic paint systems. Robotic paint systems are used for a large range of products from car bodies and parts for cars to complete aircraft. One key feature of paint robots is that they are explosion proof, as they are normally used in an environment with high solvent levels. To achieve this explosion proofing the robot arms are continually purged with clean air and the controllers are modified from the standard.

There are a number of different paint processes that can be handled by robot systems, including solvent- and water-based paints as well as two-component paints and also powder paint. Each of these requires different delivery and application equipment. Robots can handle standard air atomised spray guns, electrostatic spray guns, and also electrostatic bells. The choice of spray equipment is normally determined by the process and customer requirements. Issues such as colour change are important and robots can include banks of colour change values within the arm to provide colour change close to the spray gun, minimising the time required for the changeover as well as the wasted paint. Robots can also include full process control in that the paint flow rates, atomising air, and other process parameters can be selected and controlled within the robot programme. This provides full control of the process both for specific parts and also each individual colour.

The robots can work on both moving lines and indexing lines. Full conveyor tracking can be included to ensure the robot tracks the parts correctly as they move. Robots can also be mounted on tracks to provide for tracking of the parts through the booth. For the automotive paint applications, door, bonnet (hood) and boot (trunk) lid opening devices have also been developed to provide access to the interior areas that require painting.

Painting is one of the more complex applications for robots and requires a good understanding of the process to ensure the correct functionality is included within the robot system to achieve the results required.

3.3.3 Dispensing of Adhesives and Sealants

The application of adhesives or sealants is typically via an extrusion or a spray application. Extrusion requires contact or close proximity to the surface of the part to which the material is to be applied. Whereas a spray application provides a standoff from the part and therefore provides a larger tolerance for the path accuracy relative to the seam to which the spray is to be applied.

The materials may be single-component or two-component. The single-component material may require heating to cure, in which case an oven or similar heating device may follow the application. Two-component materials commence the curing process when they are mixed and, therefore, methods of flushing the mixed components from the system are important.

A typical system would include one (single-component) or two pumps (two-component) to provide the material and possibly some form of flow control. Temperature conditioning may also be required to ensure the optimum material properties are maintained. These devices can be integrated with the robot system, particularly the flow control, if that is beneficial for the performance of the system.

3.3.4 Cutting and Material Removal

There are many different cutting and material removal techniques, most of which have been applied to robots. Cutting techniques include sawing, routing, and water jet cutting. Circular saws have been applied to robots for sprue removal in aluminium die casting applications. Routing can be achieved using either pneumatically or electrically driven tools. Electrically driven tools are normally heavier and, therefore, require robots with higher payloads. It is most important to select the correct tool for the material to be cut and then develop the system from that point. It is also important to consider the dust produced by the cutting process, mainly for operator safety but also because in some cases the robot may require additional protection. The removal of the waste material may also be important to ensure that the operation of the robot system does not become impeded by the buildup of this waste.

Water jet cutting is a good application for robots. Water jet can provide a very clean cut, particularly for moulded or formed plastic parts. A robot can provide a 3D cutting solution, which cannot be achieved in other ways. To use a robot in this environment, additional protection may be required for the robot, due to the high level of water in the cutting booth. In addition, the feed of the water to the cutting needs to be carefully considered. The pipes feeding the cutting head cannot be flexible tubes due to the high pressure required. If the robot is carrying the cutting head, this problem is normally solved by providing a coil of stainless steel tube from the shoulder of the robot to the cutting head. This allows the robot to orientate the cutting head in three dimensions with the coil winding up or unwinding depending on the direction of movement.

Other material removal applications include polishing and deburring. These often require a specific standard of surface finish and, therefore, the process and the application of the tool are very important. As mentioned above, it is necessary to define the correct tools for the application. Deburring is often accomplished with the tools mounted on the robot and, again, either pneumatically or electrically driven tools are used. Polishing is more often achieved by presenting the parts to a series of belts, each of which have different grades of abrasive. The belts are often mounted on back stand machines, which hold and drive the belt whilst providing the correct tension. Polishing mops may also be necessary to provide a final finish, which requires the robot to hold the part and withstand the force applied by the mop as it brushes the surfaces of the part.

These applications require contact between the part and the tool; therefore, some form of compliance is often necessary to accommodate for variations in the parts or the part position. This can be achieved using pneumatics, to provide compliance within the tool, or software within the robot. However, if compliance is introduced there is likely to be variability in the results because the forces applied, or the cut made to each part, will not be exactly the same. If this variability is not acceptable, it could be necessary to include force control within the robot. This is achieved by mounting a force sensor between the tool and the robot wrist. This provides feedback of the forces applied to the robot and can be used to control the robot path to achieve the desired result.

3.4 GRIPPERS AND TOOL CHANGERS

For applications such as assembly, machine tending, and general material handling, including packing, palletising, press tending, and many other applications, the gripper is one of the most important elements of the system. Although human hand-type devices are under development they are both complex and expensive. The majority of industrial automation applications do not require this level of functionality.

Grippers are developed to suit the needs of the specific application. The method of "gripping" may use a number of alternative techniques, including the two-jaw gripper, pneumatic vacuum cups, or magnets. In some cases balloons provide a very effective mechanism.

The standard two-jaw gripper (Figure 3.7) is suitable for parts that will not be damaged by the force required to hold them. This force can be generated by pneumatics, electric drives, or hydraulics if a very high force is required. Standard gripper modules can be purchased from catalogues with

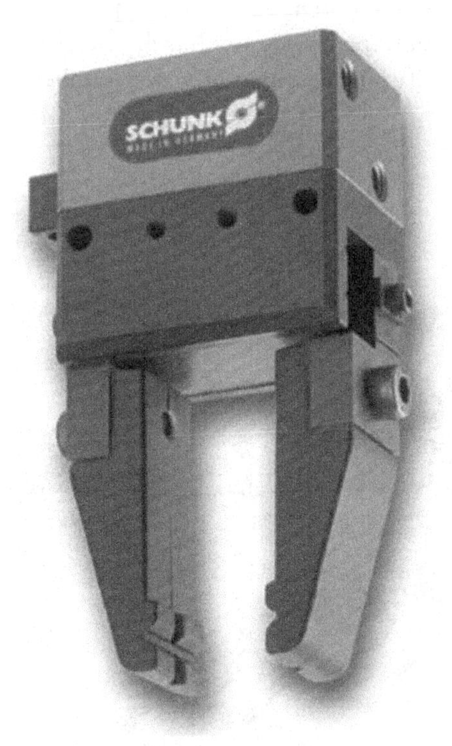

Figure 3.7 Two-jaw gripper.

the only work required being the design of the gripper jaws to suit the parts to be picked. These, or variations of these, tend to be used for machine tending applications.

Pneumatic vacuum cups are widely used for handling flat parts as well as boxes and similar objects. They are very effective, quick to operate, and normally do not damage the surface of the part being picked. The size of the vacuum cups should be kept to a minimum because larger cups take longer for the air to be evacuated from the cup impacting cycle time. The material of the cup should also be considered, particularly in relation to the temperature at which the system will operate. Rubbers that work well within vacuum grippers at 20°C can be very hard and completely ineffective at 4°C, which may be the temperature of the packing hall within which the robot system is to operate.

Vacuum grippers can also provide flexibility in that an array of cups can be included within the gripper, which is selectable to suit the parts to be picked. Redundancy can also be included by providing more cups than necessary to pick each part.

The picking of sacks for palletising is often achieved by the use of clamshell grippers (Figure 3.8). Often, a vacuum gripper cannot be used because the sacks are porous and the powder within the sack would be sucked into the vacuum system. Clamshell grippers include a number of fingers that close

Figure 3.8 Clamshell gripper.

under the sack from both sides, holding the sack underneath and, therefore, supporting this nonrigid product. To place the sack on the pallet the fingers open, basically dropping the sack onto the pallet. Although the sack is not positioned precisely, the dropping does assist in distributing the material within the sack more evenly.

Magnets are sometimes used for ferrous parts that are difficult to pick in other ways; for example, they have been used for the de-palletising of layers of unfilled tins. These would have proven very difficult to reliably pick any other way, but the magnets operated very successfully. It is worth noting that the gripper weighed much more than the parts being picked which led to the selection of a large robot to handle the required payload.

There are also more specialist grippers, such as balloons, which have been used to pick bottles. The balloon lowers into the bottle and is then inflated. This provides a very secure hold on the bottle. The handling of baked muffins is achieved by pushing a number of curved pins into the top of the muffin. These successfully hold the muffin, and the very small holes left by the pins are not visible to the consumer. Special pneumatic grippers have also been developed for picking objects with uneven surfaces, including poppadoms (thin, circular South Asian bread).

For some applications it is necessary to be able to carry multiple grippers or tools on the same robot. In some cases these can be mounted on the wrist without causing any problems of either load carrying capacity or interference with each other. In other cases this is not possible and tool changers need to be used. These are off-the-shelf products that are provided with two halves, one for the robot and one for the tool or gripper. They come in different sizes to suit different tool weights and are also able to transfer electrical power, air, and digital signals within the changer.

The actuation within the gripper is often provided by pneumatics, mainly because this minimises the weight of the gripper. Hydraulics can be used if high gripping forces are required. Electric motors are also used, including the use of servo drives when more precise control is required. However, there are cost and weight penalties when using electric drives. Hydraulics introduces a potential maintenance or reliability issue. Sensors can also be mounted on grippers to detect correct part pick up and put down as well as correct operation of elements of the gripper. Sensors can assist with the overall reliability and performance of the system; for example, by minimising the risk of mishandled parts, but they also add complexity to the gripper.

It should be noted that the gripper is the contact element of a system and, therefore, the most at risk from wear or damage. It is important that the mounting includes dowel locations to allow the gripper to be removed and replaced in a repeatable position. The gripper can then be removed for maintenance or repair and when replaced it will not be necessary to reteach any of the robot path or positions. The key to successful gripper design is to keep the gripper as simple as possible whilst still meeting the needs of the application.

3.5 TOOLING AND FIXTURING

Robots are repeatable but to provide the required results the parts to be processed also need to be located in a repeatable position. The tooling or fixturing is used to hold the parts to ensure they are correctly located in repeatable positions, which then allows the robot to perform the required operations and achieve the desired results.

Arc welding is an application where fixtures are particularly critical. The fixtures are required to hold a number of parts together to achieve repeatable positions of the weld seams and also repeatable geometry of the seam; for example, the size of the gap to be welded. However, the fixtures must also provide control of the important dimensions of the part. For example, an exhaust system consists of a number of subassemblies including bent pipes, catalyst, and silencer boxes. The most important overall dimensions are the position of the front of the system, where it would join the engine, and the position of the rear pipe, where it appears below the rear bumper. The remainder of the system can accept reasonably wide tolerances; however, for welding, the positions of the weld joints between each pipe and box are critical. The fixtures need to be able to hold the overall dimensions and, at the same time, provide repeatable positions of the seam to be welded to the robot.

In addition the fixture must provide ease of loading and unloading. One particular issue with welding fixtures is that the part will expand during welding due to the heat provided by the welding process. The part will not have cooled when it is unloaded and, therefore, will still be in an expanded state. The fixtures have to be designed to ensure the part can easily be removed even if warm after welding.

The fixtures will include tooling locations to match the geometry of the part at that position and a clamp to then hold the part in place. A number of different clamp types can be used. The simplest form is manual on/off clamps. These are low cost but are very dependent on the operator closing all clamps correctly.

The next step in cost and complexity is manual on with air backup and automatic off. These clamps are closed manually, but when the system start signal is initiated air pressure is applied to the clamps to ensure they are pushed closed; they can then be automatically opened when the cycle is complete. These clamps provide the benefit of greater certainty on the closing and reduced time for the opening, allowing the operator to unload parts more quickly.

The most expensive level of clamps is auto on and auto off. In this case, the parts are loaded into the tooling, the operator then initiates the start, and all clamps close. The main benefit is time saved both for loading and unloading. This approach also helps in terms of process reliability as the tooling needs to be a high design standard otherwise the fixture will not operate successfully. With manual clamping it is possible for the operator to make the system work even if the tooling design is poor. An automated clamping system does not provide this capability.

It is also possible to include sensors within the fixturing. These could be part-present sensors to ensure all the parts are correctly loaded. Sensors can also be included on the clamps to ensure they are operating correctly. For welding systems, the sensors need to be weld-immune to ensure reliability.

Tooling is required to hold a part in position; for example, to hold a panel for an adhesive application. Clamps are not always required for noncontact processes or when little or no force is applied to the part. For processes such as routing, a force is applied. The tooling is normally designed to the shape of the part to provide a repeatable location. Clamps or vacuum can then be used to hold the part in place and withstand the forces applied to the part during the process if this is required.

It is important to consider the material used to produce the tooling locations. If the application is welding these will normally be steel because they must be well-wearing and damage to the parts is unlikely. In some cases copper is used to help remove heat from the part and reduce the risk of weld spatter sticking to the fixture, but this will wear more quickly and will require replacement. In other applications it is important the fixture does not cause any damage to the part or the part surface and in this case engineered plastics are often used.

In many systems flexibility is required. This can require the changing of fixtures or tooling from those for one part to a new set for a different part. If this flexibility is required, the fixtures or tooling are built up on a single plate or structure. This is mounted into the robot system using doweled positions as datums. This allows a fixture or tooling plate to be removed and replaced in the same position. Following a changeover, the robot programmes that have been previously taught can then operate without any need for reteaching. It is also important in this case that the design of the fixtures provides for easy storage (e.g., on flat plates); otherwise, they can be damaged when not in use on the robot system.

It is most important that the fixtures or tooling be reliable. These are often the most bespoke elements of the robot system. Generally, simple solutions are the most reliable but additional functionality, such as sensors, can minimise the risk of any operator mistakes. It is also beneficial if poke yoke principles can be practised to ensure only the correct parts can be loaded, particularly if the same robot cell is producing a number of different variants of product. The fixturing and tooling cost can often be a significant proportion of the cost of a robot cell, particularly because this is where most of the design costs are required. Simple solutions will be lower cost but it should always be noted that any robot cell will only be as good as the parts presentation. It is therefore worthwhile investing in the tooling and fixturing to ensure the desired output is both achieved and maintained.

3.6 ASSEMBLY AUTOMATION COMPONENTS

A typical assembly system is built around a mechanism for moving the parts between the various stations at which the assembly operations take place. These operations may include adding of parts to the assembly, mechanical joining, such as riveting or screw driving, or other joining techniques, such as welding and adhesives, as well as application of seals, testing, and packaging. Testing may include electrical tests, leak tests, or visual checks, using vision, to ensure the final assembly meets the required standard and performance.

These mechanisms are often conveyors, either indexing or continuous with stop stations. The assemblies are typically held on platens that carry the fixtures onto which the parts are assembled. These are normally dedicated to the part, and a system may carry a number of the same platens/fixtures or, if a range of parts is to be produced, there will be a number of each type of platen/fixture within the system. In this latter case identification, for example, using barcodes, is required at each station to ensure the appropriate operations are performed for the type of part being presented. Each discrete operation is allocated to a specific station with the required equipment grouped at that station. To maximise the output of the system there should be enough platens within the system to ensure that no station ever waits for a platen and each station operates effectively.

Conveyor systems may take the form of a straight line, often used for indexing systems, with the initial part loading at one end and assembly off-loading at the other end. Alternatively loops can be used, typically for continuous conveyor systems. With the latter approach, it is relatively easy to include one or more manual stations within the loop to provide for part loading and unloading and also to include manual operations where the automation of these would either be too complex or too expensive.

Many of the more dedicated assembly systems are built around a rotary table, rather than a conveyor. In this case, multiple fixtures are located around the periphery of a rotary table and the equipment required for each operation is grouped at points around the table. The table would index once all the operations have been completed, moving each part onto the next stage. It is also possible to use continuously moving rotary tables with the operations being performed on the parts as they are moving. This provides for high throughput because no time is lost during indexing, but the systems are much more complex. Also it is not always feasible to perform operations on moving parts, which may be a constraint. Rotary table systems are more compact, generally lower cost than conveyor systems and can provide high throughput. However, it is not really feasible to include manual operations, and maintenance can be difficult as there is no easy access to the equipment.

An alternative approach is to build the assembly system around a robot. The robot would provide the movement of the parts through the assembly process. SCARA robots are often used for these tasks because of their high speed and compact design (see Section 2.1.2). The use of a robot provides greater flexibility because parts can be taken through different routes, depending on the requirements of the assembly process, or additional parallel stations can be more easily incorporated for operations that take the longest time. The use of the robot also provides better future proofing because it is less expensive to change the system for future product redesigns than in a more dedicated system. However, the robot approach is normally more suited for assembly operations that require a lower throughput.

Whichever route is taken for the main transport of the parts through the assembly cell, a number of other items of equipment are required to provide the functionality necessary for each operation. These may include bowl feeders or other forms of feed devices (see Section 3.1.3) to input components into the system. There may be simple pick and place devices to take the output from the feed mechanisms to place onto the assembly. There is a number of components available, ranging from single axis actuators, both pneumatic and electric, to multiaxis devices with both rotary and linear axes. These devices are equipped with grippers or carry other tools to perform the operation required. Appropriate sensors are included to provide for the sequencing of the devices and appropriate checking to ensure operations were completed successfully. A complete system could therefore be a very complex machine, generally built from standard components but with a bespoke design to suit the needs of the assembly process.

3.7 SYSTEM CONTROLS

The system control of an automation system is present to provide a number of key functions:

- Overall control of the elements of the cell or system to ensure they all operate as planned and in the correct sequence.
- Presentation of data regarding the cell or system to supervisors and operators and/or higher-level control systems such as a factorywide manufacturing execution system (MES).
- Assistance to maintenance people in the event of a fault, displaying status and error information, to provide guidance in identifying and correcting the fault.
- Overall safety functionality to ensure the cell or system can be operated and maintained in a safe manner.

The first programmable logic controller (PLC) was developed in the late 1960s with the purpose of providing this type of control. Prior to the PLC this functionality was provided by banks of hardwired relays, which were often complex, difficult to maintain and also difficult to modify for changes in the system. The PLC provided the same functionality but within software rather than banks of relays with the PLC being programmed using ladder logic, a programming method that replicated the functionality of relays and was, therefore, easy to understand and use.

PLCs are still used to provide overall system control and both their capability and programming has developed significantly. PLCs today range from microunits with only a few digital inputs and outputs (I/O) to much larger devices that can handle hundreds of I/O, analogue functions and the more advanced network interfaces such as Profibus and Ethernet (see below).

A typical robot system may well include a PLC to provide the overall control. This will also interface to a human machine interface (HMI) to provide information to the operator, maintenance, and other personnel. A typical HMI will include a screen, which is used to graphically display the status of the cell and the equipment within the cell as well as production and process information. It will also include functionality to interrogate aspects of the system, fault find and adjust important process and application parameters. The HMI will also provide the overall control, including functions such as start, stop, reset, and so on.

Within smaller robot systems, it is also feasible for the robot to provide the same functionality; therefore, a PLC may not be required. This would provide a lower-cost solution but is not always preferred because it becomes necessary for the customer's personnel to access the robot programs and can be perceived as too complicated. They may be more comfortable accessing a PLC, due to their greater experience with them; therefore, PLCs are often included to maintain a separation between the robot and its programme and the overall system control. On larger systems, for example, with multiple robots, a PLC is normally preferred as one piece of equipment which can provide the overall control rather than being dependent on the controls being distributed across a number of different elements of the system.

At the basic level the various items of equipment within the cell are interfaced to the PLC via digital I/O. Sensors are often located throughout the system to check the operating sequence and ensure a step has been completed prior to moving to the next step. These sensors might be located on conveyors to ensure parts are in position, on fixtures to ensure all parts are loaded correctly or within grippers to check parts have been picked up and put down. Robots would also provide signals to the PLC at appropriate points to indicate where they are in their programme as well as waiting for signals from the PLC before moving on to the next step.

In addition it is also possible to locate I/O blocks remotely from the host device, the PLC, or the robot controller. For example, an I/O block could be mounted on a fixture to connect to all the sensors and any actuators on the fixture. These connections may be via individual wires. The signal to the host device from the I/O block may then be transmitted via a single wire; therefore, reducing the wiring from the fixture to the host. It may reduce cost, particularly if significant distances are involved, but it does improve reliability as well as ease of maintenance and repair.

Using discrete I/O can lead to a large number of signals being routed around the cell. Both robots and PLCs are equipped with I/O blocks that typically each handle 16 inputs and 16 outputs. If a large number of signals are required this can lead to a large number of I/O blocks both within the robot and PLC and there are limits to the number of I/O blocks that can be fitted. Systems with large amounts of I/O are also more expensive to install and more complex to maintain. To alleviate this problem, networks were developed to define standards for the interfacing of equipment with the intention of ensuring compatibility between different pieces of equipment. The original industrial network was the Manufacturing Automation Protocol (MAP), originally defined by General Motors in 1982. Since that time various networks have been defined and used including Profibus and Ethernet. Unfortunately there are still some challenges as the major suppliers of PLCs tend to support a specific version of a network, reducing compatibility with the products aligned with their competitors.

Networks have also become multilayer. For example, a typical three layer network comprises:

- Device Net – a bus system that connects low-level devices directly to factory floor controllers and eliminates the hardwiring to I/O modules.
- Control Net – a higher-level network for interfacing the various machines within an automation system or on the shop floor. This may include robots, machine tools, HMIs, and PLCs.
- Ethernet – a standard information network for the fast exchange of large amounts of data between PLCs, supervisory control and data acquisition systems, as well as factorywide MES. This would provide for the overall communication and operation of automation systems across the factory floor.

Each level operates to a defined standard and products are available that meet these standards; therefore, it is possible to select the appropriate products and place these within the network knowing they will operate correctly.

The control system is also involved in the monitoring and maintenance of the safety circuits. This is discussed further in the next section. In summary, the control system provides the overall control for the automation system and also provides access to the functionality for the operator and maintenance personnel. It is therefore a key element of the system. The configuration, particularly the software, is often bespoke because the requirements of each application and customer are normally different. It is therefore important that it be designed in a logical way and fully commented and documented to allow others to understand how to interrogate the system and correct faults. Many companies now apply standard approaches to the creation of the software making it easier for others to understand and modify the code in the future.

3.8 SAFETY AND GUARDING

The primary role of the safety system is to ensure that no personnel will endanger themselves through the operation, use, or maintenance of an automation system. There are standards that explain what is required to provide this safe operating environment. Country standards may be applicable as well as standards provided by the customer. These standards are often guidelines rather than definitive statements, and require an assessment of the risk, normally by the automation supplier, of the potential human interfaces with the cell, including operators, programmers, maintenance personnel, as well as third parties. The assessment should cover all potential risks that are to be removed or minimised as appropriate by the application of the safety systems and guarding. In some cases, the assessment will be more onerous; for example, with, laser applications, due to the greater risk associated with the use of this type of equipment.

The main guarding is normally formed by a fixed guard surrounding the cell. This is normally 2 m high with a small gap allowed at the bottom. The surround is normally made from posts fixed to the floor to which are attached infill panels. The panels can be sheet metal, weld mesh, Perspex, or other forms of plastic sheet.

The choice of material is normally dependent on the application and customer's preference. For arc welding applications it is also necessary to protect personnel from the glare of the welding arc; therefore, solid panels, or weld mesh backed by weld screen material, are used. Plastic panels may well be used in cleaner environments as they provide an overall better finish to the cell but can be more easily scratched. Weld mesh is lower cost and is often used in more arduous environments where plastic panels would quickly deteriorate.

Laser applications normally require a completely light-tight box to ensure there is no possibility of transmitting the laser beam outside the cell. A recent high-power laser application provided a greater challenge in that the laser, if directed at the wall of the box, even for a very short period, would cut through the wall and therefore be transmitted outside the box. To solve this problem, the inside wall of the box is completely covered with plates, which sense the impact of the laser and cause the laser to be turned off within fractions of a second, thereby making the system safe.

Having surrounded the automation, it is then necessary to provide access. The simplest form of access is for maintenance and programming. This is infrequent and is normally accomplished by one or more access doors. These doors are interlocked to the control system so that the system cannot be operated in automatic with one of these doors open. There does need to be the provision to provide power to the machinery when access doors are open to allow the programmers or maintenance personnel to perform their required activities. There are many different solutions, but one of the most common is a key exchange system. The person enters a mechanical key into a box on the access door, which then releases a second key and allows the door to be opened. The person can then place this key into a second box inside the cell, which then allows the equipment within the cell to be operated from within the cell but not from outside. It is, therefore, under the control of the person within the cell. The cell cannot be operated from outside until the sequence has been reversed and the keys returned to their correct locations. This type of approach does require care on larger systems with multiple robots because multiple access doors may be required and multiple personnel can be within the guard at the same time. It is important that the system should not be locked with somebody still within the cell. Some companies address this by using padlocks, with individual keys allocated to each person to ensure the system cannot be closed up and returned to automatic without all personnel having exited the system.

The other area where access is normally required is for the loading and unloading of parts into the system. The requirements of the guarding for this point of entry are dependent on the equipment with which the operator comes into contact. Typical equipment used are doors, lightguards, floor mats, or scanning area guards.

Doors can be sliding, rise and fall, automatic or manually operated. These are generally built into the guard surround and provide a physical barrier between the operator and the automation when the door is shut. If the door is automatic a safety edge, that is a rubber strip that senses an obstruction, may need to be included to ensure the door cannot trap the operator in any way. When the door is open, the cell must be designed so the operator is not able to access the cell and cannot reach any moving or dangerous equipment. Doors are often used where the operator is loading to a fixture mounted on a table or a positioner. The positioner or table provides a barrier between the operator and the functioning part of the cell.

Lightguards are often used where larger areas of access are required or doors are impractical. A lightguard consists of two columns, one of which carries a series of light transmitters and the other a matching series of receivers. For a loading area, each column is mounted on either side of the area and if any of the light beams are broken, that is, the light does not reach the receiver, then the lightguard will cause a stop of the system. The output can be muted at appropriate times; for example, when it is safe to enter the area.

There are two other issues that are important. It is often necessary to determine the stopping distance of the machinery with which the operator interfaces and the distance to any trapping points. This then determines the position of the lightguard; that is, how far from the equipment the lightguards need to be placed. This is to ensure that if any personnel entered through the lightguard, the equipment would have ceased moving before the personnel reached the moving equipment.

The other issue that is important is it must not be possible for personnel to stand on the inside of the area guarded by the light guard and the system to continue to operate. This is normally addressed in two ways. First, a physical barrier is included within the cell to prevent personnel from entering through the lightguard to the load/unload area and then moving farther into the cell. The second is to either tilt the lightguards at 45° so they cover the area or to place a horizontal and vertical lightguard to ensure the area is covered. It is also feasible to use floor mats that, when stepped on, will sense the presence of the operator. However, these can be more easily damaged and become unreliable.

An alternative approach is to use scanning area guards. These guards transmit a laser beam, which is scanned across an area in front of the sensor. They then receive the return beam reflected from the objects within the range of the sensor. The sensor can be programmed to define the signal profile for a normal situation; that is, a clear load/unload area. They can, therefore, detect changes to this and cause a stop if the floor area has been entered.

It is also possible, in conjunction with software within the robot, to define specific areas within the field of the sensor. These areas are then used to cause specific actions by the robot in the event of personnel entering each of these areas. Specific robot software, such as ABB's Safemove, which meets the safety standard requirements, is required for the safe implementation of this functionality. As an illustration, two areas within the sensor view may be defined, an outer area and an inner area within the reach of the robot. If an operator enters the first area, the robot would then slow down and if the operator enters the second area it may stop in a safe mode, but not an emergency stop. Once the operator exits the relevant area the robot would return to its previous operating mode; that is, slow speed if the operator remains in the first area or full speed once the operator exits the first area. This approach can be useful for applications where interaction between the operator and the robot may be required during the robot cycle; for example, the loading of a part during the robot operation where the use of more conventional guards may either cause time delays or be impractical.

Another common guarding requirement is to provide for parts to pass into or out of a cell automatically but, at the same time, prevent personnel entry through the same area. An example of this is palletising systems where cartons enter the cell via a conveyor and both empty and full pallets need to exit the cell. The carton entry can be guarded using a tunnel constructed of fixed guards. The tunnel is built around the conveyor, large enough to permit entry of the cartons but small enough to prevent personnel access. It would also be long enough to prevent personnel from reaching into the cell.

The guarding of the pallet entry and exit is a little more complex. The pallets are typically fed in and out via a roller conveyor. Two sets of lightguards are mounted across the access point, providing an outer and inner guard. These are linked and muted at the appropriate times. For example, when the pallet is moving out of the system the inner lightguard is muted initially until the pallet has passed through, at which point it becomes active again and the outer lightguard is muted until the pallet has completely exited the cell. Therefore, the cell is guarded at all times, either via the inner or outer lightguard.

Guarding and safety is a very important aspect of all robot systems. The equipment can be dangerous if not treated appropriately and all potential risks to personnel must be addressed. Appropriate training is also required for the different types of personnel to ensure they operate the equipment correctly and also understand the safety aspects of the system. This is particularly true of maintenance and programming personnel who may well be working within the guard and are, therefore, at greater risk.

3.9 SUMMARY

The key elements of a number of typical automation systems have been outlined and some of the most important issues identified. It should be apparent that there are many different types of equipment that can be used, depending on the specific application. For each application, there are also a number of choices to be made that can have significant impact both on the cost and also the ultimate success of a project. It is, therefore, unlikely that any one person will have detailed knowledge of all the types of equipment and applications. Suppliers tend to specialise in certain applications and customers, by the nature of their business, and are also unlikely to cover more than a few applications.

It is most important when embarking on a specific automation project that good knowledge of all of the particular elements of a system is obtained, either directly or by bringing in experts as consultants or suppliers to assist the project from the initial specification through to implementation. The key to success is often not the existing knowledge or expertise but knowing what you do not know and, therefore, where assistance and advice is required.

CHAPTER 4

Typical Applications

Chapter Contents

Abstract

This chapter discusses the specific issues that should be considered for each of the main applications for industrial robots. These applications include welding, dispensing (e.g. painting), and processing (e.g. various cutting and material-removal applications), as well as handling operations such as plastic moulding, machine tool tending, palletising, picking, and packing. The discussion starts to bring together items explored in previous chapters, identifying some of the key points relating to each application. This includes the key benefits of using a robot for each these applications, as well as the specific issues that impact robot selection.

Keywords: Welding, Dispensing, Painting, Routing, Deburring, Machine tending, Palletising, Packing, Assembly

This chapter discusses the main issues facing each major robot application. Specifically, we explore the benefits related to each individual application area, as well as the key points to consider when implementing these applications. The engineer must recognise the capabilities of the end user when proposing a solution. More complex solutions increase the capability levels required for users to maintain and operate the system. Therefore, particularly for new users, it is better to keep the system as simple as possible.

4.1 WELDING

As previously noted, welding is one of the major areas of robot application, largely due to the benefits that can be obtained from the use of robots for these applications. As a result, significant development work has occurred over a number of years to provide fully integrated and reliable robot welding packages as part of standard solutions. These standard solutions are provided by both the main robot suppliers and the system integrators, and therefore, customers can choose from a large pool of suppliers.

4.1.1 Arc Welding

A typical arc welding system consists of a robot with an integrated welding package providing full control of the welding process (Figure 4.1). The parts to be welded are held in fixtures that maintain the appropriate relationship between the parts while ensuring the weld seams are presented in repeatable positions for accurate weld placement. The fixtures may be mounted in positioners to provide the opportunity to orientate the parts during the welding process, allowing access to the joints to be welded.

An arc welding cell would normally include two stations to allow the robot to weld at one station, whilst the operator unloads and reloads at the other. This arrangement maximises the utilisation of the robot and therefore provides for higher throughput. The two stations may be achieved by the use of a turntable carrying one fixture on each side or, alternatively, two individual stations. More complex systems may include multiple robots, robots on tracks, and multi-axis positioners. If it is not possible to guarantee joint position, touch sensing and through-the-arc seam tracking can be included to allow the robot to find the joints. There are also vision-based weld guidance systems that can be used.

It is important to consider the unloading and reloading operations as much as the robot welding process. In addition to being appropriate for the robot, the fixtures need to be simple for the operator. If at all possible,

Figure 4.1 Arc welding. *Source: ABB Robotics.*

the fixtures should only allow the parts to be placed in the correct location and orientation. The fixtures must also be quick to load and unload. This ensures that the robot is not waiting for the operator, which would compromise the output. Quick loading and unloading may require the use of automation in the fixtures, such as auto clamping, but there is a cost implication due to the increased sophistication of the fixtures. If the robot cell is to process a number of parts, the repeatable changeover of fixtures may be necessary, and, if required, this functionality must be built into the fixtures. Heavy parts may require the use of cranes or assisted lifting equipment to unload and reload the cell. The design of the cell must also account for this requirement, which may well have implications for the guarding design.

There is a very wide range of potential solutions ranging from the simplest (e.g. a single robot with a turntable) to multirobot systems, including multiaxis positioners and vision tracking. When considering a robot solution, the most important goal is to identify the appropriate solution preferably based on one of the many standard solutions available. As a guide, it is often better to use the simplest solution that will achieve the required weld quality and production throughput.

Users may hope to implement a complex system for processing complete assemblies because this appears to provide the best financial return. This can be difficult to achieve successfully, partly due to the control of the incoming parts but also the capability of the user to successfully operate the system. It is often better to choose a robot welding application with smaller cells producing the subassemblies. This arrangement provides two benefits. First, operating the cells requires a lower-level capability, and therefore, it provides a less challenging learning curve for the engineers and operators. Second, the output is subassemblies that are repeatable, and therefore, when the time comes to introduce the next stage (e.g. the complete assembly), the incoming parts will already be better controlled.

The main benefits of robot arc welding are consistency of performance and increased productivity. The robot places the same weld in the same place every time, using the same welding parameters, which achieves excellent results, as long as the parts and joints are presented consistently. This provides improvements in consistency and quality, reduces rework, and ensures the optimum use of energy and consumables. A typical robot system operates at 85% utilisation or higher, and it performs at this level throughout the full production period. Manual welding normally operates at about 35% utilisation, and it is subject to variation because operators cannot maintain the same level consistently throughout a shift. A typical robot system therefore produces the same output as 2 to 4 welders, dependent on the parts.

For parts such as seat frames, which require a large number of short welds, a robot can be very effective because of the speed at which it moves and reorientates between each of the welds. For larger, heavy parts, such as subassemblies for construction equipment, the speed of the robot is more constrained by the weld process. However, higher-performance processes, such as twin wire welding, can be utilised with robots, allowing the weld speed and thus, the output to be increased.

4.1.2 Spot Welding

Most spot welding robots are located in automotive body shops. These are generally complex, multirobot systems configured in cells or working on transfer lines. The main benefits of robots in these applications are the speed and repeatability of the welding. The robot places the welds exactly as required, both in terms of position and the repeatability of the weld process. Therefore, fewer welds are required to guarantee the integrity of the part being produced.

It is also possible to reduce the sizes of the weld flanges due to the positional repeatability of the robots. Robots are able to handle and orientate larger weld guns than operators, providing the ability to achieve welds that could not otherwise be made and, thus, allowing for greater flexibility in the design of parts. In this way, robots provide for greater flexibility, for a given cost, over other forms of automated welding stations, and, as a result, different models can be processed through the same facility and model design changes can be made with increased frequency.

Probably the most important issue with respect to these robot applications is the dress package providing services to the weld gun. This package must not impede the robot motion or access to the parts to be welded, but it must also be reliable. In addition, the dress package is subject to an arduous operational cycle in terms of the twisting introduced by the robot motion. Therefore, this is the item that is subjected to the most wear during operation, and, as a result, solutions have been developed that are specifically suited to the robot. It is worth noting, the cost of the dress package can be a very significant proportion of the total cost of the welding robot.

The dress package is integrated within the robot arm as far as is possible, with quick disconnect facilities at various points on the robot arm. These points are typically at the robot base, the shoulder of the robot and possibly the robot wrist. The intent is to provide the ability to quickly remove and replace a damaged element of the dress pack without the need to remove the complete package. It is often the case that an automotive body shop will have many different dress packs to suit the different application requirements across the facility. These dress packs are therefore coded so that the correct spare can be quickly identified and correctly fitted to the robot in the event of a failure. Some facilities also calibrate the robots so that, in the event of a robot problem, the complete robot arm can be removed and replaced, without the need for reprogramming, ensuring that production is quickly resumed.

Spot welding robots are also used for automotive component manufacture (Figure 4.2) and for other parts outside of the automotive industry. These tend to be much simpler systems, often with a single robot and a turntable or positioner to present the parts to the robot. The dress packs are normally subject to lower levels of wear in these cases, and therefore, they are less critical. The requirements for the design of the system are very similar to those appropriate for arc welding systems (see Section 4.1.1). The benefit of

Figure 4.2 Spot welding automotive component. *Source: ABB Robotics.*

using robots for these applications is again the speed and consistency of performance, with quality output provided at a rate that is both consistent and higher than can be achieved manually.

4.1.3 Laser Welding

Laser welding is often a costly process due to the equipment required. For many applications, manually performing the operation is not possible, and robots provide a more flexible and dextrous solution than other forms of automation. Two main issues arise with regard to the laser welding application. First, the part presentation is critical. The part fit-up needs to be precise, with minimal gaps between the parts to be welded. Otherwise, the laser process will not operate successfully. This requires the input parts to be highly repeatable with the appropriate fixturing to ensure minimal gaps. Second, the laser must be fed to the robot wrist in a way that does not compromise the robot's performance but also achieves a reliable laser feed. This is often accomplished via dedicated laser packages designed to suit the specific robot arm. There have also been laser robots developed to provide the feed integrated within the robot arm, but these tend to be very expensive, partly due to the design but also because they are currently produced in low volumes. The other very important issue is safety. Laser applications are housed within a light-tight enclosure to ensure there is no risk to operators and passing personnel. This increases the cost of the guarding package, and it also provides some constraints to the way parts can be fed into and out of the cell.

4.2 DISPENSING

Dispensing covers all forms of painting, sealing, and glue applications during which a fluid is output from a dispensing system onto the part to be processed. The robot often carries the dispensing equipment and orientates the applicator around the part to provide the coverage required. There are some cases in which the robot carries the part to a static dispensing gun, normally for adhesive applications, which provides the benefit that the robot can also perform the role of the part-handling equipment, reducing the overall cost of the cell, but this is normally limited to smaller parts.

4.2.1 Painting

Paint applications (Figure 4.3) normally consist of one or more robots painting parts mounted via a hanger on a conveyor to move the parts in front

Figure 4.3 Bumper painting. *Source: ABB Robotics.*

of the robots. The conveyor is often continuously moving because this suits the subsequent flash-off and baking processes. If the conveyor is floor mounted, it removes the risk of dirt and oil falling from the conveyor and hangers onto the painted parts. Larger parts, such as trailers for trucks, may be held stationary in the booth during the painting process, with the robots mounted on tracks to allow them to address all the areas of the parts to be painted. For smaller parts, such as enclosures for electrical items, a simpler part-handling system might be used with an operator unloading and reloading a turntable whilst feeding the painted parts into an oven.

As discussed previously (Section 3.3.2), a painting robot is normally equipped with a process package to provide the various paints and colours to the applicator, which may be a spray gun, an electrostatic gun or an electrostatic bell. The correct definition of the process equipment is a very important element of any paint application, and it is determined by the paint material, the cost benefit of the application method and other items such as the number of colours to be processed and the frequency of colour changes. It is important to consider the gun cleaning and flushing system, which may well impact the resulting paint finish if cleaning is not completed successfully. This is particularly true for two-component (2K) materials for which the mixed material will harden if not removed from the application system. As for other applications, standard packages cover a wide range of materials and application requirements. The most appropriate solution is a balance between the optimum technical solution and the cost of the solutions available.

The main benefit of a painting robot is the consistency of the application. The robot performs the same path with the defined spray parameters, which, subject to the sophistication of the system, can be tailored for different areas of the part. Therefore, the robot achieves a consistent paint film build. This is compared to manual painting that, with an expert, normally achieves the minimum film build but could also exceed this. Less experienced painters are more variable, both over- and underachieving the film build required. On the other hand, the robot solution applies the paint more efficiently, which, subject to the size of the parts, the throughput and the cost of the paint, results in significant financial savings.

This consistency also results in improved quality and a reduction in rework, again providing a financial benefit, and the removal of painters reduces the risk of dirt being introduced into the booth, improving quality and reducing rework. In addition, robots are able to achieve a higher transfer efficiency via better path consistency and control of gun triggering, meaning more of the paint will be on the part and less in the atmosphere of the booth. This not only reduces paint usage, but it also reduces booth cleaning, again saving costs. Robot applications further benefit the working environment in that operators no longer have to wear personal protective equipment (PPE) to work inside the spraying facility, not only saving the cost of the PPE, but also removing workers from this unpleasant environment.

In addition to defining the configuration of the process equipment, the designer must understand the combination of the robot reach, the size of the parts and their location on the hangers, and the time available for painting, which is governed by the speed of the conveyor. This information allows the designer to determine the number of robots needed, as well as the selection of the robot arm with the appropriate reach.

4.2.2 Adhesive and Sealant Dispensing

Sealant and adhesive applications can either use a spray or extrusion application technique (Figure 4.4), with the choice normally being determined by the material characteristics. Spray applications are often used for situations in which a cosmetic finish is not paramount or the location of the parts means the joints will be less precisely located. Extrusion requires the parts to be repeatable and the position of the joint or surface to which the material is to be applied to be positioned repeatably in all three dimensions.

Similar to the above-mentioned process applications, arc welding, and painting, the path repeatability of the dispensing robot provides the main benefit of this application, ensuring the sealant or adhesive is correctly positioned on the part. The systems may well include flow control equipment to ensure the material is dispensed in the appropriate quantity for the part. This can be integrated with the robot to vary the flow rate dependent on the requirements at various points on the part, and it can also adjust the flow rate to match the speed of the robot.

The application may also include conditioning equipment, such as temperature control, to ensure the material is applied at the optimum temperature. 2K materials may be used, and these require an appropriate mixer and also a cleaning system to remove mixed material in the event the system is not used for a period.

Figure 4.4 Glueing of head lights. *Source: ABB Robotics.*

Spray systems apply a bead of specific width. The width should be defined to ensure that the joint to be sealed is always covered after taking into account the potential variations in the position of the joint relative to the robot. The designer should account for the geometry of the joints to be processed and the size of the gaps to be filled. For example, it is very difficult to seal an outside corner because the spray tends to push the material to either side of the joint. Similarly, it is not possible to cover holes or fill large gaps because the jet pushes the material through the hole.

Extrusion systems can achieve a constant bead in the appropriate position, provided the part position does not vary. If the stand-off, the height of the nozzle above the part, varies, then the bead may become "wavy", similar to the problem encountered when icing cakes by hand. In both spray and extrusion, it is better to avoid shaped nozzles if possible. If the orientation of the nozzle is not important, it does not need to be aligned with the joint and therefore rotation is not required at corners which can increase the overall application speed.

In addition to controlling the bead, both in terms of size and position, robots normally perform the application at much higher speeds than is achievable manually. Manually, there is also the risk of "tails" of material from the end of each run, which may require cleaning. Therefore, as well as improvements in quality, the operation time is significantly reduced for a given robotic process verses a manual operation.

4.3 PROCESSING

Process applications include various methods of removing material from the parts being processed. These applications include cutting operations, as well as deburring and polishing operations. Some of the more common robot applications are discussed below.

4.3.1 Mechanical Cutting

Mechanical cutting includes sprue removal from aluminium die cast parts and plastic parts, as well as routing and trimming of plastic parts. In some cases the cutting operation is integrated within the robot machine-tending system. For example, the robot may remove the part from the die cast machine and then orientate the part to a saw to remove the sprue, before placing the completed part on an output station. Provided the cutting operation is completed within the cycle of the die cast machine, the utilisation of the robot is increased, and the cutting operation is achieved for limited additional cost.

Routing operations are normally performed on plastic parts to remove excess material from the moulding operation. This may require trimming around the part and cutting apertures within the part. The positioning of the part is important to ensure the cuts are made as required, and often the fixture matches the shape of the part and includes a vacuum to hold the part down. The selection of the cutting tool must match the material to be cut and the feed rate required.

The main criteria for robot selection are normally the reach required and the weight of the tools to be carried. It is also important to consider the force that may be applied to the robot during the cutting operation, which may require a larger robot to be specified.

4.3.2 Water Jet Cutting

Water jet cutting uses high-pressure water output from a nozzle in the form of a fine jet to provide the cutting force, and, from a robot perspective, this approach has been used very successfully for the trimming of plastic mouldings, as well as the removal of material from apertures. It is also possible to introduce an abrasive material, as a powder, into the water jet for the cutting of tougher materials such as Kevlar. This process provides very a clean cut, without producing dust, at high speeds, and it is therefore a very efficient and clean operation.

The process does produce a water mist within the atmosphere of the booth, however, and the parts must also be dried before passing to subsequent processes. This is often achieved by incorporating a drying chamber into the design of the booth. The process is also noisy and therefore, the requirement for sound attenuation must be included in the design of booth. If an abrasive water jet is used, the jet's power and cutting capability also make it more dangerous, requiring more substantial booth construction. The effect on the fixtures holding the parts must be considered as well.

There are two approaches to the process. The first is a fixed water jet head with the robot holding and orientating the part under that head. This has the advantage that the feed of the water jet can be easily achieved to the nozzle. However the gripper needs to be robust, holding the part with enough rigidity to withstand the force of the process. The robot and the gripper may well be more exposed to the process. The second approach is to hold the part in fixtures and carry the water jet nozzle on the robot. This is more of a challenge because the high-pressure water feed can only be carried in metal pipes due to the high pressure. This can reduce the orientation capability of the robot and is normally addressed by using fixed pipes to the shoulder of the robot and then a spiral of pipe wound around the upper arm to provide the feed to the nozzle mounted on the robot wrist (Figure 4.5). This spiral can wind and unwind as the robot wrist orientates and therefore does not impede the motion of the wrist. The robot should be suitably protected from the water mist, which can normally be achieved using an IP67 (see Section 2.3) rated protection to the robot arm, without requiring an additional cover over the robot.

4.3.3 Laser Cutting

Laser cutting is a highly efficient process that can be applied to robots, particularly for parts that require a 3D shape to be cut. The requirements of the

Figure 4.5 Waterjet cutting automotive bumpers. *Source: ABB Robotics.*

application are very similar to those of laser welding (see Section 4.1.3). The process is highly precise, however, in some cases, particularly for small holes, the designer should investigate the robot path performance to ensure it meets the needs of the application.

4.3.4 Grinding and Deburring

Large numbers of robots have been applied to deburring and grinding applications. This is due to the increasingly stringent health and safety standards governing these activities and the need to minimise the risk of operator "white finger" and repetitive strain injuries that often result from these processes.

There are normally two levels of application. The first is a relatively simple material-removal operation, as in the deburring of sharp edges following machining processes. The purpose of the deburring process is to protect personnel performing subsequent operations that require the handling of the parts. The application in this case does not require the achievement of any specific dimensions. Rather it is purely intended to clean the edges. This may be for engine blocks, after machining, to deburr the flash left on the surfaces that have been machined or around holes machined in the block.

A typical application may require a number of different types of tool, including cutters and belts. The robot cell may therefore require a tool change device to allow the robot to pick the relevant tool for the particular feature being deburred. The deburring tool must contact the part with a light pressure and maintain this contact throughout the path. The simplest way to achieve this consistency is often to build compliance into the tool or the mounting of the tool. It is important that the compliance mechanism is designed to ensure that the device performs as desired in all the directions required throughout the operation.

Grinding is more often associated with the removal of metal on a casting to achieve a specific dimension (Figure 4.6). This can also be the case for some deburring applications. The challenge is often caused by the variability in the amount of metal to be removed. Deburring normally requires light removal, and, if the parts are positioned securely and consistently, the robot

Figure 4.6 Milling and grinding boat propeller. *Source: ABB Robotics.*

may move the tool, without compliance, around the part on a predetermined path, and the dimensions required will be achieved. The application does need to respond to tool wear either by the regular changing of tools or a tool-wear checking system.

In most grinding applications, such as those for castings, the material to be removed is more substantial. This normally requires robots that are heavier duty and more rigid than the machines used for deburring applications, and the part holding device, either on the robot or a separate fixture, also needs to be very rigid. One approach is to programme the robot to follow the required path and to utilise a cutting tool with enough power to address the most substantial material to be removed. This has the disadvantage that the robot and cutting tool may be larger than is required for the majority of the process, and the overall speed is also constrained by the slower speed necessary for the large burrs.

As an alternative, force control devices can be incorporated. The force control device can either control the pressure or speed of the application, modifying the robot speed and path as needed to address those occasions when larger burrs are present. This approach ensures the appropriate result is achieved, while maximising the robot speed and also providing the opportunity to utilise smaller robots and cutting tools. However, force control is not a low-cost option, and it does increase the complexity of the system.

4.3.5 Polishing

Robot polishing has been applied to a wide range of components including door furniture, medical implants, and aircraft engine turbine blades. In all cases the finish is important, but in some, such as polishing turbine blades, the dimensions of the parts are equally important. The manual polishing process provides the same health and safety risks as grinding and deburring (see Section 4.3.4). However, polishing often requires the achievement of a surface finish that is much more easily attained by experienced operators and difficult to replicate within a robot system. The operators have the advantage of visual feedback as they perform the task, which currently cannot be replicated within a robot system.

For applications on items such as door furniture and medical implants, the robot often moves the parts between one or more polishing machines (Figure 4.7). There are usually a number of machines, each of which carries belts with various grades of abrasive, or polishing mops. The robot orientates the parts to these belts in turn, commencing with the coarsest and performs a

Figure 4.7 Polishing. *Source: ABB Robotics.*

sequence of moves on each grade of belt until the part is finally finished with the polishing mop.

The same approach can also be used with the smaller aero engine blades, although force control may be included to ensure the appropriate force is applied to achieve the required finish. For larger blades, these parts can be mounted in servo-driven positioners to provide orientation of the part in front of the robot. These positioners are the same as those used for arc welding applications (see Section 3.3.1). The robot carries the tool, often with force control included to govern the robot path. If a number of tools are required, then tool changing can also be built into the system.

4.4 HANDLING AND MACHINE TENDING

The largest application area for robots is handling and machine tending. This area covers a very wide range of applications in many different environments, across all different industry sectors from electronics to foundries. Even through the potential applications are broad, all designers must consider some common factors related to robot selection, which are discussed in more detail in Chapter 5. The following section explores some of the issues specific to each of the main applications within this broad area.

4.4.1 Casting

Robots have addressed various operations within casting, including unloading and die spraying for die casting machines, sand core handling and assembly, and metal pouring. The environment in these facilities, particularly in close proximity to the machines, is normally unpleasant and arduous, which provides one of the main motivations for the application of robots. In some cases, such as metal pouring, the operation is dangerous, and in others, such as sand casting, the parts are fragile, and, therefore, the handling needs to be very reliable and performed with care. In this latter case, the repeatability of the robot is very beneficial.

Die casting (Figure 4.8) is used for a wide range of parts with a wide range of sizes. If the parts must be removed from the die and carefully placed, rather

Figure 4.8 Diecasting. *Source: ABB Robotics.*

than simply ejected, a robot provides a good solution. The robot also provides the benefit of a consistent cycle time, ensuring that the die open time, which affects the cooling of the die, is consistent, thus improving the quality of the parts being produced. The same robot can also be used to spray the die lubricant, again providing consistency of application, ensuring the key areas of the die are addressed and improving quality. Alternatively, if the cycle time of the die is critical, an additional robot can be mounted on top of the die cast machine to perform the lubricant spray operation. It is also possible for one robot to address more than one die cast machine or to perform other operations during the casting cycle, such as sprue removal.

Robots used for die casting and metal pouring applications are normally equipped with higher levels of protection such as IP67, as well as an epoxy paint finish providing corrosion protection and allowing steam cleaning. This protects the robot from the harsh environment. For many die casting applications, the dies eject the part. This can cause a problem because the robot normally resists the force of the ejection. To solve this ejection challenge, a "soft servo" feature allows the servo control to be relaxed at this point. This then means the robot does not resist the force of the ejection, but rather, it moves with it, with the normal servo control being re-engaged once the part has been collected.

4.4.2 Plastic Moulding

Robots have been used for unloading injection-moulding machines (Figure 4.9) for many years. The products handled include interior and exterior automotive parts, mobile telephone covers, lawnmower parts, even widgets for beer cans. The robot tending of injection-moulding machines provides the benefit of a fixed cycle time, resulting in a fixed tool opening time and consistent die cooling, which, in turn, leads to improved quality. This has also been achieved by more dedicated automation solutions, but these are limited to the unload operation only, and they tend to drop parts to conveyors. Robots can perform secondary operations, as well as unloading, and they are also able to place the parts in the correct orientation on output, minimising the risk of damage.

The grippers tend to be simple, light structures using vacuum cups to grip the part. To ensure efficient operation, the robot must be ready to enter the tool as soon as it starts to open, and the tool must start closing as soon as the robot is clear. This maximises the throughput. Robot reach is often the key

Figure 4.9 Unloading of injection-moulding machine. *Source: ABB Robotics.*

parameter, particularly because the parts are often very light. In some cases, when very high throughput is required, the speed of operation may also be important.

Notably, the information provided on data sheets (speeds and accelerations) does not really help because these are theoretical maximums that cannot be achieved in practice. A better test is to simulate the operation to determine the actual cycle time. The robot can be interfaced to the moulding machine, as for other machine-tending applications. In the case of moulding, standards such as Euromap 12 and 67 have been created to provide a predefined interface protocol for injection-moulding machines. A standard solution results, but other interfaces can also be used.

4.4.3 Stamping and Forging

The manual operation of presses is an arduous activity often conducted in a very harsh environment, particularly in foundries. The safety implications for operators interacting with such dangerous machinery also slow the potential output of the presses due to the need to ensure personnel are clear prior to the presses cycling. The use of robots provides benefits not only by improving health and safety but also by increasing the output potential of a press facility.

Stamping operations tend to require the part to be unloaded and a new part reloaded to the press in the minimum time to maximise the output of the press. Press operations normally require more than one press with each press performing a different operation to build to the final shape. The first press receives the blank, normally from a stack, and the output from the final press is often placed in racks for shipping or moving to another area of the factory.

The tools are generally simple frames carrying vacuum cups to grip the part. The frames tend to be quite large because a long reach is often required, and they may be constructed from aluminium sections, but carbon fibre parts have also been used to improve rigidity. These tools are often attached using quick-change facilities for both the mountings and services. This is to assist the tool change that is often required on a frequent basis.

The robots tend to be mounted to one side of the area between the presses. This arrangement provides access for die carts to the presses to allow the dies to be changed quickly. The reach requirement then determines the main criteria for the choice of robot and the size of the gripper frames. The robots can also be interfaced to the presses to minimise the cycle time. This can be achieved by mounting encoders on the press movement. The robot can then be programmed so that it starts to enter the press before the die is fully open, and the die begins closing before the robot has fully removed the gripper and part.

Press forging also requires parts to be moved between presses, although the parts are very hot and can even be over 1000°C. In addition, the environment contains oil mist from the die lubricant spray. The robots therefore require protection from the environment, such as that described in Section 4.4.1. The robots are typically equipped with extended tools to grip the part, while keeping the robot wrist outside the area of the forging press. The grippers also include heat protection, and, in some cases, hydraulically powered grippers may be required to provide the force required to hold the billets.

4.4.4 Machine Tool Tending

The main benefit of applying robots to machine tools arises from the removal of manual intervention with the machine. Machine tools are already automated, and the only manual operation required is the unloading of the completed part and the loading of the new part to be processed (Figure 4.10). This manual intervention can cause delays to the machining cycle if the operator is not ready to perform the changeover as soon as the

Figure 4.10 Machine tool tending. *Source: ABB Robotics.*

machine is finished or does not perform the changeover as quickly as possible. The machine also stops when unattended, as happens during breaks.

Typically, a robot is fitted with a double gripper so it is already holding the new part when it enters the machine. It can then quickly remove the finished part and reload the machine without having to exit the machine. The changeover is therefore accomplished faster than can be achieved manually.

As a result, the automation of the load process does improve the utilisation of the machine, increasing output. It is also possible to provide buffers to allow the robot to continue feeding the machine during breaks and outside of normal production hours. This may lead to a reduction in the number of machines required for a specific production rate, as well as the space required to house those machines. If the machining cycle is lengthy, it may also be possible for the robot to perform secondary operations, such as deburring, during the time available providing additional benefits. Alternatively, if a number of machines are required, either to perform different operations on the same part, or to provide parallel machining activities, the robot may be able to work with multiple machines, increasing the utilisation of the robot and improving the effectiveness of the automation.

One example of a current standard system includes a robot, input and output conveyors, and a vision system. The input conveyor is loaded with parts that do not need to be precisely located. The vision system then identifies the position of the part for the robot, which can then pick the part and load it to the machine tool. Once a part is completed, it is loaded to the output conveyor. As a variation, bins of parts can be loaded into the system with conveyors automatically feeding parts from the bin to the robot. These systems can therefore be loaded with a batch of parts and run unattended for a number of hours.

The savings produced by the automation do not purely result from labour saving. They also result from the increased output from existing machines and, therefore, reduced investment in additional machines. These additional savings can be significant and may form a very important element of any justification.

The selection of a suitable robot is mainly driven by the reach and weight-carrying capacity required. The reach is determined by the size of and access to the machine, as well as the other elements of the system around the robot. Normally, floor-mounted robots are the preferred route, but in some cases, they may be mounted overhead to improve the access. Robots may also be mounted on tracks to provide the opportunity to service multiple machines. Again these tracks can be mounted on gantries, providing the benefit of maintaining a relatively clear floor area and therefore easy manual access to the machines for tooling changeovers and maintenance.

With regard to weight-carrying capacity, the designer must consider the weight of the gripper and the location of the part, when held in the gripper, relative to the wrist of the robot. This offset can be significant and therefore reduces the weight the robot can carry (see Section 2.2). If a double gripper is used, the robot carries two parts for a period within the cycle, and this additional weight must be accounted for.

To provide for an effective system, the designer must also consider issues related to the machine, such as the removal of the swarf resulting from the machining operation, as well as the cutting-tool life and the frequency of tool changeovers. If these are frequent, the operation might benefit from automatic tool changing within the machine tool. If the machine tool is to produce a range of parts, access for tooling changeovers must be considered, in addition to possible gripper changes on the robot and access for maintenance of the machine tool.

Robots may also be included as an element of the operation for the machine, as happens with pipe-bending machines and press brakes

Figure 4.11 Tending of press brake. *Source: ABB Robotics.*

(Figure 4.11). In these cases, the robot provides orientation of the part between operations to ensure the correct changes in form are achieved. These systems provide similar benefits to machine tool-tending applications, particularly relating to increased productivity, part repeatability and reduced health and safety issues.

4.4.5 Measurement, Inspection, and Testing

Robots are used for many inspection applications. The main benefit is the ability to move the measurement devices, such as vision systems, to a number of different points to inspect or measure different parts of the product to be tested. Robots have also been applied to mechanical test rigs to provide motion as an alternative to bespoke motion systems. In both cases, the benefit of using a robot is the reduction in cost of the overall system. In the former case, a smaller number of measurement devices are required and in the latter the cost of the robot may be lower than the cost of designing and manufacturing the bespoke system.

One key issue when considering the use of robots is to remember that these machines are not accurate, and, therefore, the designer must compensate for their inherent inaccuracy. If absolute measurements are being made, then appropriate compensation or reference measurements must also be made. In addition, it may be necessary to include temperature compensation to account for physical changes to the robot due to changes in the environment.

4.4.6 Palletising

Palletising is a major robot application due to the speed of operation and flexibility a robot can offer, as well as the reliability provided by a robot system. A robot system provides benefits, in comparison with manual operations, based on labour saving, increased speed of operation and also improved health and safety, by the removal of lifting activities. In comparison with more dedicated palletising machines, a robot provides increased flexibility and, in some cases, particularly for light packages, improved palletising reliability. The increased flexibility lets the operation palletise different packages to different pallets, even if they arrive on the same line, reducing the time lost due to pallet changeovers.

Normally, a four-axis robot is used for palletising applications because it is not necessary to change the orientation of the package (Figure 4.12). The choice of robot is largely based on weight capacity and reach, and the designer determines it based partly on the product to be palletised and the layout. The required cycle time may also influence the design if it is necessary to pick multiple packages or even complete layers for each cycle.

Figure 4.12 Palletising paint buckets. *Source: ABB Robotics.*

Section 5.2.3 further discusses the selection of the robot in relation to concept development.

The gripping method is an important element of the system. In many cases, a pneumatic solution is both viable and generally preferred due to simplicity, speed of operation, gripper weight, and cost. This can be achieved via an array of individual pneumatic cups or, alternatively, a Unigripper-type approach. The Unigripper normally provides more flexibility than a simple array of vacuum cups. For sacks, particularly of powders and loose materials, a clam shell-type gripper is often used (see Section 3.4). These are very effective for this type of product, but they are slower than vacuum grippers. It is also possible to use mechanical grippers, such as fork lift-type devices or even plates, to grip the part. These place constraints on the way a stack can be built on the pallet, because they do require access on the sides of the package and, therefore, are slower than a vacuum device. Their use needs to be carefully considered to ensure they do not damage the package, but, in some cases, they are the optimum approach.

4.4.7 Packing and Picking

The benefits of applying robots to packing and picking applications are often throughput; labour saving, particularly for multishift operations; and consistency leading to improved quality. Vision systems are often required to identify the location of the products to be picked, and, as a secondary feature, they can provide quality control. In some cases, particularly where the products are heavy, there are also health and safety benefits.

If very high speed is required and the products are light, the system might benefit most from inverted delta-type robots (Figure 4.13), such as the FlexPicker (see Section 2.1.4). These are specifically designed for this type of application, and although they are generally four-axis machines, this axis arrangement provides the motion required for most picking and packing applications. In some applications, particularly secondary packing applications during which the products are placed into boxes or cartons, six-axis robots might be more effective because they provide additional orientation capabilities. Articulated robots are also a better configuration for reaching into boxes.

For many primary packing applications, the gripping technique is the challenge, and the weight of the gripper may well be higher than the weight of the product. Therefore, the choice of robot cannot be determined until the basic design of the gripper is known. Also, these primary packing

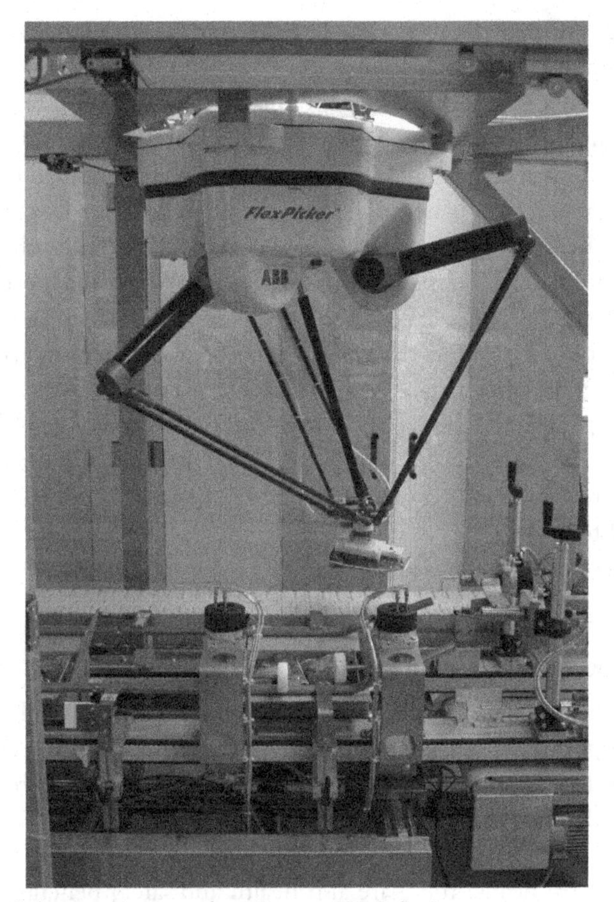

Figure 4.13 Robot packing pouches. *Source: ABB Robotics.*

applications are often performed in high care areas of food factories where hygiene is critical, and the robots must meet the relevant standards, as well as being designed for washdown cleaning (see Section 2.3).

4.5 ASSEMBLY

Assembly is a growing application area, with the electronics industry being a major user, especially in the area of consumer electronic products. A typical robot assembly system is shown in Figure 4.14. Robots provide precision, speed, and consistency that are not achievable manually, as well as flexibility that cannot be provided by more dedicated assembly systems. Typically SCARA robots (see Section 2.1.2) have been used for assembly, but six-axis

Figure 4.14 Robot assembly system. *Source: ABB Robotics.*

and delta robots are increasingly employed in this application. Delta robots can achieve high speed, but they have a limited working envelope and must be mounted above the area in which they are to operate. SCARA robots are also fast, and, like deltas, they are limited to four axes, but they can provide a very effective working area. SCARA machines tend to be lower cost than both the equivalent delta and six-axis machines. The main benefit of six-axis machines is their flexibility, which provides additional orientation capability and reduces the risk in the assembly system design.

The most important parameter for most assembly applications is speed, particularly for electronics assembly. Specifications for robots in assembly applications include the standard goal post test (see Section 2.1.2). This gives a time for standard type of assembly move up, across and down, and, therefore, it provides a way of comparing different robots and configurations for this particular move. Payload can be important and must account for the weight of the gripper mounted on the robot. Reach and size may also be important, often because the assembly systems need to be compact. As a result, a compact robot may be better suited than a larger machine.

In addition to the above parameters, mechanical assembly applications may require a force to press two parts together. In this case, the ability of the robot to apply that force may be important. This type of information is not normally apparent from the standard data sheets, and it may require a more-detailed discussion with the suppliers. Assembly systems may also include joining functions, using additional fixings such as screws or bolts. It is possible to automate many of these operations, but some are much easier than others. The design of the product and the fixing method also significantly impacts the ease of automation and the reliability of the final solution.

One of the most important issues for assembly applications is positional repeatability. If two parts are to be joined by some mechanical method, the parts must be located within tolerances that allow the joining technique to operate. This may require some form of lead in or chamfer on the parts. It may require vision to identify part locations. It is necessary for the parts themselves to be repeatable. For example, if components are to be inserted to a printed circuit board, the legs on the components must be straight and undamaged. One of the other major challenges for assembly is the feeding of components (see Section 3.1.3). The use of bowl feeders and other mechanisms to reliably feed the parts to the assembly system is key element of the success of the system.

CHAPTER 5

Developing a Solution

Chapter Contents

Abstract

This chapter outlines the steps required to develop the concept for a potential automation application. The initial step is the definition of the key parameters, such as the part details and production rates. We then discuss the development of the initial concept in terms of the main applications, including arc welding, machine tool tending, palletising, packing, and assembly, before providing information for other applications. We also identify the main control and safety issues considered at the concept stage, as well as the use of testing and simulation in refining the initial concept.

Keywords: Arc welding, Assembly, Cycle time, Dispensing, Machine tool tending, Painting, Primary packing, Secondary packing, Kinematic simulation, Discrete event simulation

Irrespective of application or complexity, every robot system should be developed using a number of steps and iterations in order to achieve the optimal outcome. The goal should always be the on-time implementation of a successful system within budget. A successful system is one that achieves the desired production rate and quality objectives with the anticipated financial return. Many factors can impact the financial return, and we discuss these factors further in Chapter 7.

This chapter illustrates the steps required to achieve a successful conclusion to the project, highlighting the main pitfalls and how these can be avoided. Nothing about a robot system makes its development and implementation significantly different from any major capital project, other than the lack of awareness about what a robot can do and, more importantly, what it cannot, especially in relation to human workers. Robots do not have the same intelligence and sensory abilities as manual labour, and, therefore, they cannot be a direct replacement for human workers. The design of the solution needs to suit the capabilities of the robot system in relation to the parts and the process. This can, and often does, require changes to the manual process in order to suit a robotic process, but the overall result is often an improvement.

It is also worth noting that the introduction of a new robot to replace an older robot is a much simpler proposition. In this case, the majority of the process and application are already defined, although a detailed study may identify how the performance of the new machine could provide opportunities for improved output or quality from the system. The decision process should also consider the reuse of the existing equipment or the replacement of this equipment, as well as the robot. The older equipment may compromise the performance of the new robot, and, therefore, the short-term savings may be offset by increased expenditure in the longer term. It is also necessary to consider safety because standards often change from the date when the original system was installed. The guarding and safety system may need to be modified and updated to meet the current standards that will apply due to the introduction of the new robot. Although a robot replacement does appear to be straightforward, a detailed investigation is required. Without appropriate study, the cost of a robot replacement can quickly escalate, leading to problems with the agreed-upon budget.

5.1 DETERMINING APPLICATION PARAMETERS

The first step toward the development of a successful project is to obtain a detailed understanding of the current application. This may seem obvious, but details can be missed at this stage, which may not become apparent until commissioning, and such oversights can have a very significant impact on the success or cost of a project. The following discussion is based on an ideal situation because all the information is often not available. However, engineers should obtain as much of the important information as possible, while also developing an understanding of the risk related to any unknowns.

First, the engineer should obtain all relevant drawings and documentation for the parts to be produced, as well as the basic process details. For example, if considering a welding project, the engineer must ascertain the type of welding process, size and type of wire, welding gas, weld sizes, and specification. It can often be beneficial to view the actual parts and the current operation if possible. One of the most important issues to identify is the variability of the parts, as presented to this operation. This could include dimensional variances, variability of surface finish, cleanliness, and part presentation. The key is to understand the reality rather than what may be defined on a drawing or within a specification. An understanding of the important parameters is also useful as this will vary for different applications, processes, and solutions.

For example, for a palletising application, the security of the carton-closing method is important if vacuum is used to pick the boxes. For a welding application, the processes and machines used to prepare the subassemblies are important because they provide a guide to the repeatability of the parts and, therefore, the repeatability of the weld joints. In this case laser-cutting the parts, followed by bending on modern press brakes, is likely to provide repeatable parts that can be robot-welded. Part dimensional variances and the cleanliness of the parts may be particularly important for an assembly system. The method of presenting parts to a machine loading application may be important. It is very difficult to provide an all-encompassing list because the parameters depend on the process to be performed, the types of parts and the proposed automation solution. Further details are discussed in Chapter 6, which covers the development of a specification. If this is done correctly, all these points are addressed.

If possible, the engineer should also discuss the application with the person currently performing the process. Operators are very flexible and generally do solve problems. They may be doing something within their activity that makes the job either possible or easier, but this adaptation might not be detailed on any of the process documentation or work plans. For example, the operator may make a change to accommodate input variability that is not identified on the relevant documentation. The operator may also solve part fit-up problems or other problems may have been identified and solved by production without informing other parts of the organisation. Without knowledge of these activities, the engineer may devise an automated solution that costs more than anticipated due to the additional work or equipment required to achieve a working system, or the design may actually fail to perform as required.

The next step is to define the required production rate, number of shifts the equipment is to be used for, number of hours per shift, and number of working weeks per year. The objective is to determine the cycle time required to achieve the necessary production output. During this stage, the engineer also tries to establish the anticipated efficiency with which the robot cell is expected to operate. This includes downtime due to breakdowns, planned maintenance, and replenishment of consumables, as well as the downtime of other facilities, which could affect the uptime of the robot cell. This other downtime may include lunch breaks and other stoppage times if the cell is not able to operate without operator support. Based on this information, the engineer can define a target cycle time that will achieve the desired production rate at the anticipated efficiency. This is basically:

$$\text{Target cycle time} = \text{total time available}/\text{required output} * \text{efficiency factor}$$

The efficiency factor used for these calculations may well be as low as 85%. The mean time between failures for a typical robot suggests that the efficiency should be in the high 90s, but the efficiency factors for each element of the system multiply together, which reduces the overall efficiency, and it is normally better to be conservative, particularly during the early stages of the project. In addition to the cycle time, the engineer must understand the required operations on the parts and the relevant process parameters. For an arc welding application, this includes the location of the welds, the weld sizes and lengths, details of the weld joints, the physical access to the welds, and the size and weight of the parts.

5.2 INITIAL CONCEPT DESIGN

The initial concept design often results from previous experience as much as it does from a detailed study of the application. Having previously viewed similar applications, the engineer may be able to suggest an outline concept as a possible solution. The concept design is an iterative process, however, and having an outline concept as a starting point can be very helpful. The thought process by which this initial concept is developed can depend on the application being considered. The following sections provide some guidance for concept development and the main issues related to designing a solution for the main robot applications. These applications are arc welding, machine tool tending, palletising, packing, and assembly. We also provide more general comments on other applications, including spot and laser welding, painting and dispensing, as well as material removal.

5.2.1 Arc Welding

The simplest arc welding system consists of a robot and welding package working on two tables. The fixtures to locate the parts are mounted on these tables. Two tables are included to allow the operator to unload and reload one table whilst the robot is welding at the other. Provided the robot cycle time is longer than the operator unload and load times, the robot works continuously, and the output of the robot is maximised.

An alternative design might include a manual turntable in place of the two fixed tables. This potentially has the benefit of reducing the need for operator movement, because the worker does not need to walk between the two fixed stations. It also simplifies the guarding and potentially improves the robot access to the part, particularly from the sides, which are completely open in the case of a turntable. The turntable may be beneficial if the robot is only producing one part at a time. In other words, the two fixtures are the same. Therefore, the operator can load from the same bins and also unload the completed parts to the same place. If different parts are being produced, two fixed tables may be more effective because these provide two distinct stations, reducing the congestion around one station and the risk of mixing parts. The time to change between stations is also reduced because the manual turntable depends on the operator to make the turn, whereas the robot can switch between fixed stations as soon as it is ready.

A slightly more complex system includes a powered turntable. This removes the dependence on the operator for the turning, and it may be required if heavier or larger parts are involved. It does provide improved access to the parts in the same way as a manual turntable, but it complicates the guarding because the turntable also needs to be guarded, increasing the size of the cell. The choice between fixed table, manual turntables, and powered turntables can often result from preference and a perception as to which is more appropriate for a particular facility.

In designing an arc welding system, the engineer must initially consider the requirement for orientation of the part during the welding process. It may be necessary to position the weld joint in a preferred orientation (e.g. 45°) to optimise the weld process and, particularly, to avoid welding vertically upwards. In both such cases, a positioner may be used to provide the optimum conditions for welding rather than access for the robot. In some cases it is possible to include a two-stage process with parts welded in one orientation in the first stage and then in a different orientation in the second stage. The operator can turn the parts between the stages. For example, the operator could use a fixed table with two fixture locations. During the load cycle, the operator removes the fully welded part from the second location, unloads the part from the first location, reorientates the part and places it into the second location and then loads new parts into the first location. Therefore, for each cycle, a new part is loaded, and a fully welded part is produced. It is also be feasible for additional parts to be loaded into the second fixture location if required.

This orientation capability may be required because a weld needs to be made around a tube or a box. It may be preferable to perform this weld in one pass to avoid starts and stops in the weld path. For example, if the weld is to be leak proof, starts and stops should be minimised. Therefore, to provide access all the way round the part, the part must be reorientated during the weld process. If part reorientation is required during the weld process, or it is not possible to manually reorientate parts between cycles (e.g. if the parts are too heavy or the reorientation is too frequent), then a positioner may be required in the cell. The basic positioner consists of a headstock carrying a single servo-driven axis and a tailstock (Figure 3.3). The fixture is typically located between the head and tailstocks and can therefore be rotated using the servo-controlled axis under control of the robot programme. The most important issues to be considered are the weight and size of the parts. These single-axis positioners are available in a range of sizes according to the weight capacity, the maximum swing diameter, and the maximum length. The weight capacity must be enough to accommodate both the weight of the part and the fixture, which is often heavier than the part. The swing diameter must be large enough to accommodate the maximum diameter of the part and fixture combined, and the length must allow the fixture to fit between the head- and tailstocks.

If these single-axis positioners are required, the engineer often includes two positioners, to provide for unloading and reloading at one whilst the robot welds at the other, in the same way as two fixed tables. These can be positioned on either side of the robot, but this arrangement does complicate the guarding because suitable screening must be included to protect the operator from arc light being produced at the operating station. Locating the positioners on either side does separate the two stations, however, which can be beneficial if different parts are being processed, reducing congestion at the load stations and also potentially improving the flow of parts around the cell. But it also potentially increases the travel distance for the operator if he or she is working on both stations.

An alternative approach is to place the two positioners side by side, which simplifies the guarding but requires the robot to be mounted on a servo-driven track to ensure it has the reach needed to access both positioners, increasing the cost. For longer parts, the use of a track may be necessary anyway, so this may not be a major implication for the solution.

There are also two station versions of these single-axis positioners (Figure 3.4). In effect, these are two single-axis positioners mounted on a turntable. The same issues, weight capacity and size, apply regarding the selection of the most appropriate as for the single axis positioners. The swing radius of the turntable must also be considered to ensure the robot is positioned outside this swing radius. This often means the robot has to be located further from the parts to be welded, and, therefore, a larger robot may be required. The swing radius also governs the maximum length of part that can be accommodated; as a result these two-station positioners are only appropriate for smaller parts, otherwise the swing radius becomes impractical. Alternatively, the robot can be mounted overhead so that it is away from the swing of the turntable, but this arrangement introduces other issues, particularly regarding maintenance.

A further type is the H-style positioner (Figure 5.1). This is similar because it carries two single-axis positioners, but instead of rotating around a vertical axis, the index rotates around a horizontal axis. Therefore, the robot can be positioned close to the welding station. It is also possible to accommodate longer parts because the length is not constrained by the swing radius. Both these types of two-station positioners include a weld screen between the stations that protects the operator from the arc glare produced as the robot is welding.

There are also two-axis positioners that can rotate the part in two directions (Figure 3.5). These can orientate a specific weld to exactly the orientation required, and they are typically used with subassemblies for off-road vehicles, often consisting of thicker metal parts and requiring multipass welding. These two-axis positioners are also available as two-station devices with two mounted either side of a turntable. This positioner is therefore be a four-axis system.

The cost of the positioners can add significant cost to the cell. In addition, the fixtures can be expensive items. If the production rate and cycle time are

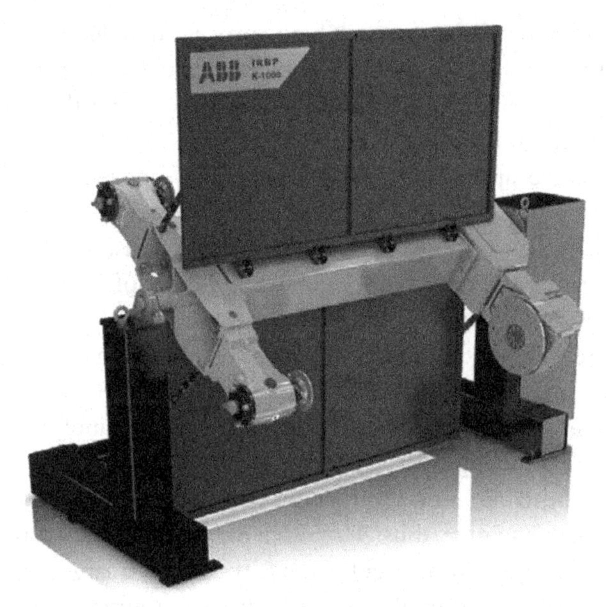

Figure 5.1 H-style two-station positioner. *Source: ABB Robotics.*

such that more than one robot is required, it is often beneficial to include two robots within one cell. This may remove a requirement for a track, because the two robots can cover a larger area, and it is obviously less expensive than two separate systems. If two robots are included, the work load between the robots must be balanced in order to ensure that the output of the cell is maximised.

At this stage, the engineer has probably not designed the fixtures, but, based on experience, the engineer can often estimate the weight and size for the parts to be welded. The type of positioner and solution design may have some influence on the fixture design, and therefore, the engineer must define a basic concept before the fixtures can be designed.

As mentioned above, the engineer must consider how the guarding can be designed to ensure that the system can operate safely. It may also be necessary to consider how the parts will be unloaded and reloaded to the system. Some parts can be heavy, requiring assisted lifting or overhead cranes, and access needs to be provided to the load/unload stations for this equipment. It may be necessary to consider the loading on the positioner during the load/unload process. Two-station positioners might be affected by the physical loading of very heavy parts. If vibration of the positioner is created at the load station, it can be transmitted to the weld station, resulting in potential problems with the welds. This should be considered as part of the concept stage and may suggest the use of single-station positioners.

It is also important to consider the flow of parts through the robot cell, both from and to adjacent operations. This is partly to minimise congestion, to ensure that the part feeding is not a problem, and also to help the operators achieve required tasks without difficulty. To this end, the design can include a combination of fixed tables, turntables, and positioners to provide what is required. Normally, a system contains a minimum of two stations, but, in some cases, more may be beneficial. For example, if frequent fixture changes are required, three or four stations may be included. With suitable guarding, this allows the robot and operator to work on two stations whilst fixture changeovers are being performed on the others.

Once the basic layout is defined, including the selection of positioner, the engineer should consider the type of welding equipment. First, the engineer defines the weld process package needed to meet the requirements of the application. The main robot issue at this stage is the weight of the welding equipment. For example, twin-wire welding equipment requires a much heavier torch than single-wire systems. This establishes the required weight capacity of the robot. The reach required of the robot is based on the parts and the part presentation, or the choice of positioner. Using this information, the engineer can determine the optimum robot for the concept. As part of the consideration, the mounting of the robot should be assessed. The preference is for floor mounting, but inverted mounting can be appropriate. If inverted is the optimum, this may affect robot selection, because not all robots can be mounted in an inverted position. Finally, the engineer should define the locations of the welding power source, wire feed, and weld torch service centre.

Although there are many different selections to be made in the development of a welding solution, the constraints of the parts and production rates normally define the type of solution that is most appropriate. The type of positioner can often be determined from experience, and this then drives the rest of the solution design.

5.2.2 Machine Tool Tending

In many ways machine tending is much simpler than arc welding because there are not the same choices to be made in regard to associated equipment such as positioners. The first important issue in creating a machine tool tending application is the weight and design of the parts to be handled. The objective is often to maximise the output of the machine tool. Therefore, the time taken for unloading and reloading should be minimised. To achieve this, robots are often equipped with a double gripper so that they can remove the finished part and replace it with a new part without losing the time needed to withdraw from the machine, deposit the finished part, and collect a new part. As a result, the robot often needs to be able to carry both a finished part and a new part, as well as the dual gripper. The aperture within which the unloading/loading operation takes place (see below) may also be important. With regard to part weights, the new part weighs more than a finished part because the machining process removes some of the material of the part and therefore reduces the weight. The loading on the robot is also often offset from the centre of the wrist axis, particularly when it is only carrying one part. In addition, the size of the gripper is important because this determines where the centre of gravity of the part will be relative to the wrist. As discussed in Section 2.2, the load capacity of the robot is both a function of the weight of the part and the position of the centre of gravity with respect to the mounting flange on the wrist. In some cases, the gripping method is different for a new part and a finished part due to the operations that have been performed within the machine tool. In this case, a double gripper is still required, even if only one part is being handled at any one time.

Once the load capacity has been determined, the engineer must consider the reach. The reach is partly determined by the access into the machine tool, both in terms of the reach required to access the fixture on the machine bed or the jaws of the chuck, and the size and location of the aperture provided for loading. Not only does this have an impact on the reach of the robot, but it may also affect the design of the gripper and its ability to hold two parts. If the parts are relatively large and/or the aperture is small, the robot may not be able to access the machine with a dual gripper and orientate this gripper inside the machine whilst carrying both parts. If this is the case, the robot needs to remove one part before placing a new part into the machine. To minimise the effect of this movement on machining time, the engineer may include some buffer locations just outside the machine tool to provide temporary storage for parts. As part of this consideration, the automation of the door opening may be an issue. With some machines, the opening for an automated door is less than that of the manual door. Therefore, this detail needs to be checked and may become important.

As part of the robot reach study, the engineer must also consider the other stations or activities that the robot will address during its cycle. The parts may be presented on pallets or fed from a conveyor or stack. The finished parts may be returned to the same pallets, stacked on different pallets, or even fed out on a conveyor. These input and output stations often include some form of buffering to reduce the dependence on operator intervention. If the buffering is large enough, the system may be able to operate for extended hours without any manual intervention. This could extend to overnight or full days on Saturday and Sunday. Therefore, additional capacity can be achieved through the machine tool, without the need for personnel to be present. The main constraints to the success of this approach are the reliability of the equipment and process and the size of buffer required.

Two or more machine tools are often addressed by the same robot, with the robot taking parts from one machine and feeding them to the next. Each of these stations needs to be grouped around the robot in a logical sequence in order to minimise the time required by the robot to move between each station. The overall sizes and access also need to be determined, and from them, the engineer can define the reach of the robot. In some cases, particularly if larger machine tools are involved, the robot may be mounted on a servo-driven track to provide the overall reach required.

In some cases the robot also performs further operations, such as deburring, within its cycle. This can be achieved via a tool mounted on a stand, with the robot presenting the part to the tool whilst holding it within the gripper. Alternatively, the robot may present the part to an inspection station prior to loading it to the output device. If the robot has time available, it can engage in additional operations, built in at a relatively low cost, thus improving the overall justification for the system. The engineer must consider all of these issues prior to robot selection.

Due to other constraints within the system or the differences in the parts to be handled, the design may include tool changers. These provide the ability to automatically change grippers (see Section 3.4). There is time required for the tool change, but the system's overall flexibility is significantly enhanced. The use of tool changers requires a rack for the various different grippers to be used, and the rack needs to be located in an appropriate position within the cell.

It may be necessary to include multiple robots within the handling system to achieve the throughput required. In this case handover stations can allow one robot to pass a part to another robot. If two or more robots are required, a better solution may be to split the systems. This has the advantage of simplifying each system, but it may increase the space required and the cost. The overall objective should be to maximise the output of the machine tools, achieving the product rate required whilst minimising the investment in the automation.

The design should always provide access to the machine tools for maintenance or fixture changeovers. Both of these functions are likely to be necessary at some time. The only point of access may well be the same aperture used by the robot to unload and reload parts. Therefore, the cell design needs to ensure that the appropriate access can be achieved, and this access provision might require assisted lifting equipment or overhead cranes for heavier parts. In some cases, if the robot is tending multiple machine tools, the guarding may have to allow manual access to an individual machine without causing the rest of the system to stop. In other words, the guarding may need to allow manual access to one machine whilst at the same time preventing robot access to that machine and also preventing manual access to other areas of the system.

The effectiveness of a machine tool-tending application normally depends on the need to unload and reload the machine as fast as possible to ensure the machine operates at the highest efficiency possible. The robot is the tool that provides this capability. It may be possible to include additional machines or operations within the robot cycle, provided the robot has time available. It may also be possible to operate unmanned working conditions, given appropriately buffered infeed and outfeed. This can increase the capacity of the machine tools for limited additional costs, and it can remove the need for further investment in additional machine tools.

5.2.3 Palletising

Palletising normally involves picking boxes from an input conveyor and placing them onto a pallet. In most cases a four-axis robot is used because it is not necessary to reorientate the boxes other than by rotating them about the vertical axis, and therefore, four axes are all that is required. The four-axis solution provides a larger carrying capacity than the equivalent six-axis machine, as well as a lower price. Both features principally result from the absence of the fifth and sixth axes of the wrist.

In most cases, the selection of the robot is reasonably simple because it is determined by the weight capacity and the reach required. The reach is partly driven by the size of the pallet, although most palletising robots can palletise to the standard sizes of pallet. When determining reach, the engineer should focus on the height of the full pallet to ensure the robot can achieve a full layer at this stack height. The maximum is normally 2 m, although in many cases the stack height is lower than this.

The weight capacity requirement is driven partly by the weight of each individual object (box, sack, or other item) to be palletised, the gripping method and the number of each item being picked. The number of each item being picked is normally determined by the throughput required. To achieve the throughput, the system may palletise two or more items or even a complete layer within one cycle. This has implications on the gripper and the feeding system. The simplest solution involves one box arriving at a pick station, located via a pusher to ensure it is consistently positioned, picked by the robot, and placed on the stack on a pallet. The gripper often utilises pneumatics to provide a quick and simple gripping technique. The same approach can also be used with other objects such as sacks being picked via a clam shell gripper (Figure 3.7).

If cycle time requirements result in the robot needing to pick more than one item at a time, the system often becomes more complicated. For example, with two items being picked, the part infeed system must collate the two items in a consistent position, and the gripper needs to be able to pick the two items simultaneously. In most cases the items are also placed simultaneously, although, in some cases, they may need to be placed individually. This approach can be extended to multiple items and eventually a full layer. As more items become involved, the choice of grippers is basically reduced to pneumatics only. Sometimes, due to the nature of the parts, the palletising operation is more reliable if complete layers are formed, picked, and palletised. However, if many items are to be picked in one move it is necessary for the part infeed to collate these items in a reliable and consistent format.

The infeed can therefore require careful consideration, particularly if more than one product is being fed into the system. It is quite feasible for a robot to palletise a number of different products to different pallets. The only constraint is the number of pallets the robot can reach. Normally, the maximum around any individual robot is four, but this can be extended if the robot is placed on a servo-driven track. This does increase the palletising time, however. It may be better and probably about the same cost to use two robots rather than a robot mounted on a track. The infeed can be down the same conveyor with some form of product identifier, such as bar code readers, identifying the pallet onto which the robot is to place the item. Alternatively, a robot can handle more than one input conveyor, with each input normally dedicated to one output pallet station.

The number of output stations can also vary. As mentioned above, a typical palletising robot can handle four stations unless the working envelope is extended by mounting the robot on a track. There are two issues that the engineer must consider regarding the number of output stations. The first is the number of unique pallets required. Most palletising systems place boxes from one input onto one pallet. Normally, only distribution and logistics systems require larger numbers of output pallets. The most significant issue with most palletising systems is the changeover time when a pallet is full. If only one item is being palletised to one pallet, it is feasible to have two output stations. The robot can automatically changeover to the second station and continue to palletise. At some appropriate point, an operator can arrive, remove the full pallet, and load a new pallet ready for the robot. This simple solution can be extended to four pallet stations with two input conveyors and the robot palletising two different items, one from each input, to one of two pallet stations. Providing that the operator removes the full pallet and replaces it with an empty pallet within the time taken for the robot to complete a pallet, the system continues to run without interruption.

For systems requiring frequent pallet changes or systems that do not rely on operator interaction, there are two different approaches. The first involves a pallet stack, loaded by the operator, with, say, 10 pallets, but unloaded by the robot, taking a pallet when required and placing it in the load station. The second approach is to use an automated pallet dispenser. Again, the operator loads the stack, but a new pallet is dispensed and fed via conveyor to the load station. The first approach does require some robot time to transfer the empty pallet, but the second is completely independent of the robot, and therefore, it has no impact on the palletising cycle. Both of these approaches require the full pallets to be removed from the system. The best way of reducing the system's dependence on the operator is to include conveyors to drive the completed pallets out of the loading station. This provides a buffer, and, as long as the full pallets are removed before the next pallet is completed, there is no delay to the palletising. If this may be an issue, the buffers can be extended. It is also possible to automatically feed the completed pallets to an automated shrink wrap machine that often follows palletising.

The requirement for layer sheets also needs to be considered. These sheets are normally cardboard, and they are provided from a stack that is replenished manually. The stack needs to be within the working area of the robot, and the robot gripper needs to include the capability to pick up these sheets from the stack and place them onto the pallet stack.

One issue to address during the concepting of a palletising system is the guarding. As mentioned in Section 3.8, it is important to ensure there is no opportunity for operators to access different areas of the palletising cell. Two stage guards are often used on the output for the full pallets to ensure the pallets can exit to the area accessible by the operator for unloading, but operators are prevented from entering the cell.

Similar to other applications, the most important issue for a palletising system is achieving the required throughput. The gripping mechanism can be a challenge for palletising, dependent on the type of item to be palletised and the consistency of the presentation or packaging. The other major issue can often be space. Palletising systems can use quite a large area due to the sizes of the pallets, the conveying required, and the need to ensure a safe perimeter. System design must therefore ensure the concept fits within the available space.

5.2.4 Packing

Packing involves placing product into its point-of-sale packaging, followed by the packing of this into transport packaging. This process could be primary packing in which, for example, meat or bread products are being packaged into their primary packing. This is often plastic or polystyrene trays that are followed by a shrink wrapping system. A secondary packing application requires numbers of these trays to be placed into larger plastic trays or cardboard cartons for transport to the customer. Other products, such as cosmetics, do not require primary packing because the products are already contained in cans or bottles. A further example, which has often been automated, is the packing of chocolates into boxes, the primary packaging, followed by the packing of these boxes into transport packaging, larger boxes. Having packaged such items into the secondary packaging, these automation systems often feed the packages to palletising systems in which the trays or cartons are placed on pallets.

The first step in developing a packing system concept is to determine the throughput required, the type of product, and how it will be input to the packaging system. It is also important to identify the type of packaging required from the output of the automation system.

Primary Packing

Products are normally presented to an automation system on conveyors, and they are often randomly distributed because they are fed from previous operations for which the control of product positions is not required. A robot system must then pick these items from the conveyor and place them into the primary packaging. If the primary packaging is plastic trays or some similar container, these are often fed from magazines and loaded onto a conveyor. This conveyor usually has flights to push the trays along, ensuring their position is known and controlled. The tray conveyor normally runs parallel with the product conveyor. This minimises the distance between the pick position and the placement position, thereby optimising the time required for transfer.

The robot is equipped with a gripper to pick the product from the conveyor. The design of the gripper needs to be appropriate for the product to be picked. It must not damage the product, must pick and place reliably, and must also be fast. The gripper is often pneumatic because this mechanism provides for quick pick up and put down. There are products, such as muffins, for which vacuum cups are not appropriate, and special devices based on curved pins have been developed to actually grip the product. These do leave small holes but they are not visible to the final customer. Chocolates are often picked with delicate mechanical grippers to avoid damaging the top of the chocolate. Vacuum solutions have also been developed that do not contact the surface of the product, and even irregular and delicate objects such as poppadums can be picked successfully. Simple vacuum cups provide the quickest method of gripping and also the lightest grippers, and, as a result, they are the preferred solution. The engineer must always remember that these picking robots often achieve high accelerations, and the product must be held as the robot moves between the pick and placement positions.

As mentioned previously, the products are often randomly located on the input conveyor. In these cases, a vision system identifies the position of the products to guide the robot to pick correctly. The vision system is located at the entry point of the conveyor into the robot system. Conveyor tracking also allows the robots to track the product as they move down the conveyor. The tray conveyor may also be tracked to ensure the positions of the trays are known in the event that the tray conveyor slows or stops, although this is not always required.

The first step in designing a packing system is to determine the number and type of robots required, based on the required throughput, the weight of the products, and the anticipated pick rate. Typically, delta robots are used for these applications because they provide high acceleration and therefore high pick rates. They are limited in terms of weight capacity, however, which may place constraints on the gripper design. Normally delta robots also only have four axes, and thus, they cannot change the orientation of the product, other than by rotation, between the pick and placement positions. If the product orientation needs to be changed, the engineer must select a six-axis machine.

It may be appropriate for the robot to pick and place one product at a time, or, alternatively, it may be possible to pick a number of products and place the group into the tray. The feasibility of the latter approach depends on the weight capacity of the robot, the weight and size of the gripper, and also any collation pattern required in the tray. Therefore, the engineer must find a balance between the requirements of the application and the capabilities of the robot type selected. It is also important to consider the environment into which the equipment is to be placed. If it is a food application, such as a high-care area, the design must meet hygiene specifications. Such specifications have an impact on the selection of the robots because not all robots are manufactured to meet these standards.

Once the type and number of robots have been defined, the engineer can determine the layout. As mentioned previously, the tray conveyor normally runs parallel to the product conveyor, with the vision system located above the product conveyor at the input to the cell. If they are the Delta configuration, the robots are mounted above the product conveyor positioned to provide the reach required to access both the picking area and the placement position. There is typically a line of robots, each working on an area of the product conveyor and a corresponding area of tray conveyor. The output from the robot system is trays of product that are usuaally fed to a shrink wrap or flow wrap machine. The tray conveyor therefore needs to interface with this wrapping machine, and this is often incorporated within the automation system.

Secondary Packing

Secondary packing applications are much more varied. As mentioned above, these can follow a primary packing application so that the product has already been placed into a pack and these packs are to be collated into boxes or trays suitable for shipment. Alternatively, individual cans, jars, bottles or tins can be placed into packages for shipment. In some cases these packs form the point-of-sale packaging, as with a cardboard tray placed on the supermarket shelf so that the consumer can pick the individual items from it. The secondary packaging can therefore be a sealed box, a plastic tray or a cardboard tray with a shrink wrap cover.

The input is often, but not always, more controlled than it is for a primary packing application. The first step is to consider the throughput, the type of product to be packed, and the packaging into which it is to be placed. The type of product heavily influences the gripping method, with vacuum often preferred for speed, although mechanical grippers may be required. To achieve higher throughputs with the minimum numbers of robots, it may be preferable to pick more than one product at a time. When placing bottles, cans, and similar objects into a cardboard box or tray, it is often easier to place one complete layer at one time. This provides for a more reliable placement because there is no risk of incorrectly positioned product getting in the way of subsequent placements. The type of placement pattern may also have an impact. For example, packing plastic trays into a box may require some to be turned in order to allow the maximum quantity to be fitted into the box. This may require a different number of product to be picked during each picking cycle to match the requirements of the placement.

The placement requirements define the collation of the products at the pick position. It may be beneficial to group the product in the positions required for the placement to allow the group of products to be easily placed in one move. This is normally achieved mechanically via the infeed conveyor because this provides a quick and reliable solution. The robot can reorientate the product if required, particularly if it is a simple matter of rotation. It is also possible to modify the grouping as part of the gripper functionality. The simplest change that can be accommodated a modification of the spacing between the products. For example, a conveyor may provide a group of products that are spaced apart, but the gripper can then close up the spacing prior to placement in the packaging.

Once the number of products to be picked has been defined, the weight of the gripper can be estimated. The weight of the product is known, and therefore, the load capacity of the robot can be determined. The need for reorientation of the product is also known from the placement patterns, and using this information, the engineer can select the appropriate robot type. Based on the throughput required and the number of products picked for each cycle, the number of robots can then be determined. Normally, one robot would work on one pick station and one put-down station, and if multiple robots are required, multiple stations are also necessary.

As part of the system concept, the engineer must determine the forming and feeding of the packaging. Plastic trays are normally fed from a magazine. The important issue is the frequency with which the magazine needs to be replenished. If cardboard trays or boxes are to be used, these likely require forming. This can be achieved using dedicated tray or box formers integrated within the automation system. These formers are very suitable for higher-volume applications, but they may be overly expensive for lower-volume applications. For these latter applications, a robot combined with some dedicated tooling can form the boxes or trays, with the card being fed from a magazine. The approach of using a robot may be appropriate, particularly if greater flexibility is required, and the robot can also perform other tasks within the system.

The requirement for layer sheets should also be considered. In packaging these can often be sheets of paper provided from a dispenser. Typically, the dispenser holds a roll that it outputs, cutting the sheet as required. The robot needs to be able grip the sheet and place it reliably into the box.

Once the tray has been filled, it is normally output to a stack. If a cardboard tray is used, it is normally shrink or flow wrapped, and thus, the completed tray has to be output to a wrapping machine. If the output is a box, it is normally closed and sealed using either tape or glue. All these operations can be accomplished by relatively standard equipment. Alternatively, a robot with the necessary equipment and tooling can be used, provided there is time available. This can be particularly beneficial if a more flexible solution is required. There is often the requirement to place a label on the pack, which again can be accomplished via an automated labeller or via a robot. The use of a robot for these additional operations (box forming, sealing, and labelling) can sometimes be combined with the palletising. The decision as to which route is most appropriate is very dependent on the flexibility required and the target throughput. The key is to provide a balanced solution that achieves what is required with all elements being used effectively.

5.2.5 Assembly

There are typically two forms of assembly application. The first is the addition of new components to a product, and the second is the application of some fixing method, which may include dispensing glues or adding screws or other mechanical fixing techniques. The first consideration in designing such systems is the number of operations required and which are to be automated. It is often beneficial to avoid the automation of the most complex operations because these pose the greatest risk. They are also often the most expensive to automate and can make the complete solution unviable financially.

Therefore, the engineer must break down the assembly into a series of operations or tasks, each of which must be completed in sequence. There may be a number of subassemblies that are built in parallel prior to bringing them together to form the final assembly. The number of operations tends to define the type of assembly system. If the number is small, it may be feasible to include these within a simple cell. For more complex assembly tasks requiring a number of stages, some form of part transport, such as a conveyor or rotary table, may be appropriate to carry the product through the assembly stages.

Once this is understood, the engineer can determine the automation equipment required for each stage. It is also important to consider the part-feeding equipment and the feeding of the fixing devices. As part of this development process, the engineer estimates the time required for each stage. It is important that this is balanced because the overall output is determined by the slowest stage. Some operations may need to be broken down to provide fast cycle times or multiple stations provided in parallel to achieve the desired cycle time. Similarly, if certain stages are under-utilised, the engineer may be able to combine operations, leading to an overall reduction in cost.

The determination of the equipment for each stage influences the handling method. If significant amounts of equipment are required, a rotary table may not be appropriate due to the space constraint it imposes. Similarly, if manual operations are also required within the sequence, it is easier to include these operations through the use of individual cells or a conveyor that moves the parts between the operations.

The flexibility required also has an impact, particularly in terms of the number and/or complexity of the fixtures, grippers, and tooling. Within conveyor-based systems, it is relatively easy to have a number of platens carrying different fixtures, each dedicated to a product type, being processed through an automation system. This approach also provides the greatest flexibility for future product changes. The most flexible approach includes a number of individual cells, each processing one or more stages of the assembly. Such designs require greater manual involvement, however, and they also require some method of part transfer between cells. They are therefore less efficient.

Within each cell or assembly stage, the operations required define the equipment to be included. Within a typical assembly system, the main part is located via a fixture, either on a platen on a conveyor or mounted on a table, either fixed or rotary. The operation to be performed on this main part, be it the addition of further parts or the addition of some fixings, is typically performed by a robot or other mechanical device that brings the new parts or the tools to the main part. One aspect of assembly that often differs from this approach is the testing stage during which the assembly may be loaded into a test fixture by a robot.

As with all robot applications, the engineer should always consider the equipment to be used, the parts being assembled, and the ease and reliability with which the operation can be achieved. The best approach may be to take the part from its fixture to the tools or to place it in a temporary fixture, whilst the operation is performed. For example, an assembly may have one operation during which the part needs to be turned over. It is often better to build this capability into the station where the turning is required rather than into every fixture. The removal of a part to a new station also provides the opportunity to include multiple stations if parallel operations are required to achieve throughput or if the product variants require different operations or fixtures.

5.2.6 Other Applications

The process for determining initial concepts for other applications follows thought processes that are very similar to those discussed above.

Spot and Laser Welding

Designing a laser welding application is very similar to designing an arc welding system, although safety is a more significant issue. Spot welding may also be very similar, although, particularly with subassemblies, it may be preferable to hold the parts in the robot gripper and present them to one or more weld guns or pedestal welders. This approach can often allow the use of smaller robots, because the parts are lower weight than the weld guns. This approach reduces component handling because the robot carries the parts, often picking from and returning them to a tooling plate. It is also possible for the operator to load fixtures, which are then gripped by the robot in order to present the parts to the welding equipment. On completion of the welding, the fixture is returned to the load/unload station to allow the operator to remove the completed part and to load new parts to the fixture.

Painting and Dispensing

General dispensing applications can also be very similar, with the key decision being whether to carry the part or the dispensing equipment. Painting applications do require different approaches largely because the robots are normally explosion proof due the presence of solvents. In most painting applications, the part is presented to a robot carrying the spray equipment. The part may be presented via a continuous or indexing conveyor or in some cases via a simple device to move the parts to be painted into the spray booth. In simple terms, the choice of robot is largely dependent on the size of the part to be painted. However, a painting application is very specialised and does require specific process knowledge to ensure that the correct robot package, including the process equipment, is chosen for the application.

Material Removal

Cutting processes, such as water jet and routing, normally require the engineer to consider robot reach and access to the areas of the parts to be processed. Water jet systems do require the robot to be protected from the environment, as well as a specific dress package if the robot is to carry the cutting head. Processes that apply a force, such as routing and deburring, also require the engineer to account for the effects of that force when selecting a robot. In these cases, the engineer must often decide whether the robot carries the part or the tool. This decision is influenced by the weight of the part, the method by which the part arrives and the access required for the process. If the part does not weigh too much, a smaller robot can carry it, and, if the part requires reorientation during the process, a better solution might involve the robot carrying the part. In this case, the robot can also handle the part infeed and outfeed, which can reduce the cost of the cell.

For simple deburring tasks, with the single goal of removing a burr left by a machining process, a compliant tool is used to ensure that the required contact force is applied to the part by the tool, and therefore, the burr is removed. In some cases, force control can be included within the robot to ensure the correct force is applied during the process. The force control device is normally mounted between the robot wrist and the gripper or tool, and it provides feedback to the robot about the forces that occur as the process is performed. This additional control can ensure that the desired result is achieved and the process is performed effectively.

5.3 CONTROLS AND SAFETY

Chapter 3 provides some discussion of controls and safety. The approach for safety is very dependent on the legislation appropriate for a particular country, as well as any factory-specific requirements. These standards do change on a regular basis, and it is important that system designers both know and follow the necessary legislation. Similarly, many factories have standards in relation to controls and particularly human machine interfaces (HMI) that they expect to be implemented on their automation systems. It is therefore not possible to provide a comprehensive guide.

The main requirement for safety is that any operators, maintenance personnel, or other personnel in the factory cannot be placed in danger. This generally means that the system must be guarded in a way that prevents access into the working area of the automation, except in controlled circumstances. The safety and guarding must also prevent the automation from causing problems outside the guarded area. For example, if there is a risk of a part being dropped due to a gripper failure, the system must retain the part within the guarded area. This is normally achieved via a guard surrounding the cell. This guard is typically 2m high, consisting of posts and infill panels. The infill is typically sheet metal, weld mesh, or a plastic sheet, such as makralon or polycarbonate. Sheet metal is used where it is necessary to protect personnel from arc glare from a welding process. Weld mesh can be used for many applications. On the other hand, some customers prefer clear plastic because it provides a good cosmetic finish, as well as good visibility of the automation within the cell.

There are two forms of access required for an automation system. The first is for the loading and unloading of parts, and the second is for maintenance operations or programming activities. Any operator interaction with a robot cell, as might occur during the loading of parts, must be provided with a load area that allows access at the appropriate times whilst also preventing access further into the cell. This area must be guarded in such a way that the automation recognises if inappropriate access has been made to the load area. This recognition must then cause the automation operating in that area to stop. For example, a welding cell might include a turntable that the operator unloads and reloads. This can be guarded using a light guard. The light guard allows access when the turntable is stationary for the operator to perform the necessary tasks. Once the operator has exited the area and indicated the cycle can continue, normally via a pushbutton, the light guard senses any subsequent entry and immediately removes power from the turntable, causing it to stop. The robot system may also be included in this stop, but if the area of the robot is protected via additional guarding, the robot may continue to operate.

For maintenance purposes, access doors are normally provided at different points in the guarding. The number of doors depends on the size of the system and the elements of equipment within the system. These doors are sensed by the safety system, and if they are opened, the system causes the equipment to stop. The doors normally incorporate a key system that allows the equipment to be operated in a safe mode with personnel inside the guarding perimeter. If possible equipment that requires regular maintenance or replenishment is placed alongside the guard walls, with access provided for the necessary activity, so that it can be safely achieved from outside the guard perimeter.

Safety is most important and should be considered early in the development of a concept to ensure the system can be operated effectively and in a safe manner. The overall control and integration of all elements of a system is often achieved using a programmable logic controller (PLC). The PLC also drives the HMI through which the operator and maintenance personnel interact with the system. The HMI is particularly important in the event of stoppages because it can provide an indication of what has failed, thereby quickly directing the maintenance personnel to the problem. In terms of the concept design, the main issue to be considered is the location of the control panel, housing the PLC, and also the number and location of the HMI.

5.4 TESTING AND SIMULATION

As part of the concept development, the engineer may need to test the proposed solution to provide the confidence that the desired result is achievable. This may involve an actual test to prove the process or a simulation to demonstrate robot reach and cycle time. If a test is to be performed, it should be as realistic as possible. For example, if testing a proposed welding application, using the recommended type of welding equipment to weld the parts to be produced provides a good indication of the feasibility of the application. If positioners are proposed, it may not be feasible to use the actual positioner, but an alternative may be available. Unless the item to be produced by the system already exists, the proposed fixtures cannot easily be tested, but normally, a suitable trial can be performed, which, based on the experience of those involved, provides a very good indication of the results. Generally, applications such as welding are tested to show the end customer that the application is viable.

A number of applications involve processes with a lower level of proof. For example, for water jet cutting and routing, the designer may perform a trial to determine the cutting speed that can be used. Other material removal applications may be tested in the same way to provide a guide for the cycle time. For handling applications, particularly those involving the handling of unusual products, a trial may offer proof of principle that the gripper technique will be successful. This is particularly true for primary packing in the food industry when the products to be picked have not been handled by robots previously. This type of trial contributes to gripper development. It can therefore be a lengthy exercise requiring a number of iterations before a successful gripper concept is defined. The investment is worthwhile, however, because expensive mistakes can be avoided.

The engineer can use simulation as an alternative or addition to the above types of tests. It provides the benefit that a more detailed study can be undertaken for a relatively low cost. Two types of simulation may be appropriate. The first is kinematic simulation of the robot and associated equipment. The second is a discrete event simulation that provides a model of the operation of the facility, including any automation to check for production rates and resource requirements. Standard packages exist for both types of simulation, and automation system suppliers or end users can use them to develop and test concepts and solutions.

Kinematic simulation provides the opportunity to create a model of the robot cell. The robots can be input to the model from libraries, as can other standard components, such as weld torches, positioners, and grippers. The 3D CAD models of the parts to be processed can also be input, and any special devices can be created. The robot or robots can then be programmed to perform the task and the complete cycle simulated. This allows robot reach to be checked, as well as any potential collisions within the cell. The simulation also provides a good estimate of the cycle time. The simulation can potentially be used to check a number of different robot types and also to compare alternative solutions. Finally, the guarding and all other equipment can be added to provide a full 3D model of the system. This can be very useful when seeking approval from senior management because it provides a good visual representation of the system as it is operating. The simulation can also output the robot programme, which reduces programming time during the installation and commissioning phases of the project.

Discrete event simulation is more of a black box approach, with each black box being a specific operation within a facility. The simulation models the inputs, outputs, cycle times, and efficiencies of each operation, as well as any resource requirements and interactions with other operations. Discrete event simulation is not performed for individual robot cells. It provides a model of the complete facility and can be used to identify bottlenecks, the effect of changing operations or resources, the effect of introducing new equipment or the impact of downtime in a particular operation. The simulation provides the opportunity to check alternative solutions, equipment, and resources. The operator can run "what if" scenarios that can provide very useful input into the design stage, particularly for larger or more complex facilities.

Both tools can add value to the concept design phase. In particular, kinematic simulation tools are now used extensively by automation system suppliers to develop concepts because the simulation provides information that cannot be easily obtained in other ways. This approach therefore reduces the risk in a project.

5.5 REFINING THE CONCEPT

After developing an initial concept based on the requirements of the process, the parts to be addressed and the production rate required, the engineer must often further refine the concept before the optimum solution is determined. There are typically a number of iterations of the initial concept to ensure the final concept provides a workable and cost-effective automation solution for the specific application. A number of issues should be considered before the concept is finalised. These issues are discussed below.

First, the level of flexibility within the system should be considered. This relates not just to the current requirements but also to the future use of the system. There may be a need to vary production rates and batch sizes, as well as changes in the product itself, and the engineer may be able to develop a very flexible concept, particularly if it is largely based around robots. However, this may incur costs that are unnecessary. When built in at the start, flexibility requires a certain investment, but the cost of changes later will be lower. If more dedicated equipment is used, the initial cost may be lower, but the cost of changes later could well be high. It is therefore important to strike a balance between the cost and the flexibility.

With regard to the flexibility required, the changeover time between products should also be considered. For example, it may be necessary to change fixtures within a welding system, or grippers may need to be changed in a handling system. If product changes are frequent, say multiple times within one shift, the system may benefit from automated changeover, such as automatic tool changers for grippers, or the inclusion of equipment and facilities that provide for quick fixture changeovers. If batch sizes are large and changeovers are therefore less frequent, it may be acceptable to adopt a more manual approach. The former has cost implications, whereas a manual approach, although cheaper, requires a longer changeover time.

The engineer should carefully consider the number of robots and their utilisation. In general terms, it is better to ensure each robot is fully utilised, but there may be occasions within multiple robots systems that a robot is included to perform a specific task but is otherwise not fully utilised. It may be possible to include other tasks within the workload of this machine, provided this does not overly complicate the system, by requiring additional equipment or otherwise compromising other aspects of the system. On some occasions a robot is included because, although it is under-utilised, the flexibility provided and the reduced risk, through the use of standard equipment, means that an overall better solution is achieved. Where multiple robots are performing the same or very similar tasks, the engineer must ensure the workload is balanced between the robots.

The input of the product to the system and the output from the system must also be considered. If manual input or output is required, it must be achieved safely both in terms of interaction with the automation but also in terms of the weight and sizes of the product to be handled. For example, a welding system for car exhausts can be manually loaded because the components are small. However, the unloading presents a challenge because the output is a fully welded exhaust that is both heavy and unwieldy. In this case, a robot unloading the welding fixture to a conveyor may be appropriate. If heavy parts are to be manually loaded and unloaded, there must be access for lifting equipment. If a large number of parts are to be input, there must be the room for appropriate storage of the input parts. The cycle time of the automation may well achieve the throughput required, but the time required for operator loading and unloading must also be considered to ensure that this does not detrimentally affect the throughput. If the system requires the replenishment of magazines or other feed mechanisms at different points, then manual access needs to be provided, preferably without the need to stop the automation during this replenishment.

The engineer should also consider the effects of downtime, in the forms of unplanned and planned maintenance. The anticipated reliability of each stage of the automated process should be assessed to identify those stages that are most at risk of a problem. If a particular operation is likely to require more maintenance, then access to address this problem must be provided. Access needs to be quick and safe. Otherwise, the downtime will be extended. For example, access hatches for the tip dressing of spot weld guns do not require the operator to enter the cell and therefore minimise downtime.

The space available within the factory for the location of the automation must be addressed. This can impact the overall system design, particularly when the various operator access points, for part input and output as well as maintenance, need to be considered. It may be necessary to consider not only the floor space but also the available height and any existing roof supports or other equipment that cannot be moved. The system layout and configuration can sometimes be compromised by issues outside the system itself, and these factors need to be addressed at an early stage in the concept design.

The overall efficiency of the system should be considered as well as the efficiency of the individual elements within a system. Based on these efficiencies, will the target throughput be achieved? If not, what can be done to address this? If a particular operation is likely to be less efficient than the rest of the system, will this reduce the throughout below the target? If this is the case, can this operation be decoupled from the rest of the system and appropriate buffers included to ensure downtime at this operation does not impact the overall system efficiency? In some cases, based on this efficiency assessment, the engineer may need to include additional robots and equipment to ensure the throughput can be achieved. Alternatively, it may be possible to avoid additional equipment if buffers can be included at the input and output to allow the system to catch up in the event of downtime, particularly if the requirement is marginal.

Normally, the most important issue is the cost of the system. The concept design needs to take account of the cost savings because the project will not go ahead if an appropriate return on the investment cannot be achieved (see Chapter 7). As mentioned above, the flexibility built into the system often affects the cost. Also, the automation of some stages, although technically feasible, may prove cost prohibitive, and they should therefore remain as manual operations. In many cases the optimum solution is reasonably obvious, and there is little choice as to which route to take. However, particularly in the more complex systems, alternatives do exist, and one choice may be more cost effective than another. If the solution appears to be cost prohibitive, it may be worth investigating further to determine if these costs can be addressed. In looking at the source of the costs in an arc welding system, for example, the engineer might identify equipment included to provide compensation for the variability of components being input to the system. It may be possible to reduce this variability by modifying operations prior to the automation, thereby reducing the cost of the automation. For a packing or machine tending application, costs may be required to handle the positional variability of the product arriving at the robot system. Modification of existing equipment or changes to manual procedures may solve this problem with little additional expenditure.

The development of an automation concept is normally an iterative process with improvements being made at each stage. It is often beneficial to involve others in this process to ensure that the optimum solution is obtained and to ensure that all the potential issues are addressed. This approach can also address issues beyond the automation system, such as the layout and the input and output of product. It is often worthwhile to look beyond the confines of the operations to be automated in order to ensure nothing is included that could be simply addressed elsewhere. Overall, the engineer should strive to keep the design simple. A simple approach is likely to be the most cost effective, the most reliable, and the easiest to implement and operate.

CHAPTER 6

Specification Preparation

Chapter Contents

Abstract

This chapter describes how a specification conveys the needs of the customer while providing a common basis for comparison of supplier proposals. To this end, we discuss the main items covered by the specification, including the requirements for the automation solution (e.g. parts to be processed and production rates) and the customer requirements for the execution of the project (e.g. project management expectations and project timing). We also address the key issue of buy-off criteria, together with the testing stages that should be performed within a project.

Keywords: User requirements specification, Factory acceptance tests, Site acceptance tests, Turn-key, Buy-off, Golden part

The purpose of a specification is twofold. First, the specification conveys to potential vendors the requirements for the automation solution. In other words, it presents information such as the production rates, parts to be handled, and any standards to be applied. Second, the specification defines how the project is to be handled. This includes the time scale, any reporting requirements, and most importantly, the testing that will define if the system has met the requirements of the customer. As a result, the specification is often referred to as the user requirements specification (URS).

Most automation systems are purchased without a detailed specification. Many customers rely on verbal conversations with potential vendors to define their wishes. The vendor then provides a quotation that may include some of these requirements, but the quotation is often phrased in a way that suits the vendor. Some requirements may have been misinterpreted by the vendor, or even missed completely. There are two problems with this approach.

First, it is very difficult for the customer to compare quotations from different vendors. They may nominally be quoting to the same objective, but there may be significant differences between the quotations that are not apparent on paper. Second, the lack of a real specification is only acceptable until the system fails to meet the customer requirements or until problems are encountered in the execution of the project. Although it is better for all projects to be executed as anticipated, with the results always meeting the expectations of the customer, this outcome is less likely if the vendor is not using a clearly defined specification. Without the specification, something is more likely to go wrong. In many ways this is the fault of the customer: 'If you don't tell them what you want, then don't be surprised if you don't get it'. The development of a detailed specification can be time consuming, but it is very worthwhile, both in terms of vendor selection and project execution. There are many ways to convey this information, and the level of detail can vary significantly. The size and scope of the automation system can also impact the size of the specification. However, there are a number of specific elements that should be included, and these are reviewed in the following sections.

6.1 FUNCTIONAL ELEMENTS OF A SPECIFICATION

As mentioned above, the first objective of the specification is to convey to potential vendors the requirements for the automation project. To achieve this goal, a number of key issues need to be addressed.

6.1.1 Overview

The first step is to provide an outline of the current operation and process. The intent is to clarify the context for the automation project. This outline should include the products being addressed, the operations performed, and the manual input required. Any significant issues with the current operation should be highlighted, particularly if the automation system is intended to offer a solution to these issues.

6.1.2 Automation Concept

Based on the overview, the specification then explains the purpose of the automation system. This part introduces the specific requirements to be achieved by the automation, although these will be specified in more detail later. At this stage, the specification should highlight the key stages in the automation system, including the part input and output and how this would integrate with the existing operations. Any initial thoughts as to potential concepts and solutions should be identified to provide guidance to the potential vendors. Although the customer may not have developed the final solution and may also be open to alternative approaches, the specification should offer some guidance because this assists the vendors in understanding the objectives of the customer and the level of automation they will be considering. Providing this information also helps vendors determine if the project is suitable for their expertise. Providing clear direction reduces the time spent by the vendors in the development of their proposals. It also reduces the time required by the customer in both providing the assistance required by the vendors to understand the requirements and the assessment of the proposals when they arrive.

6.1.3 Requirements

This section of the specification is critical because it defines the main parameters and operating requirements for the automation system. First, details of the products are provided. These details include dimensions, weights, and possibly drawings for each of the parts and any assemblies to be considered. The specification must state any tolerances on these dimensions, as well as any tolerances on the presentation of the input parts, if relevant. These tolerances must be the actual tolerances that the automation must handle rather than anything detailed on drawings. Therefore, this section can provide the major challenge. However, it is key to the success of the project. If the system is developed to accommodate a range of tolerances but, in reality, it is presented with parts outside of those tolerances, it will not operate successfully.

The output requirements must also be specified. In particular, the document should address whether the output is to be presented in a specific way (e.g. stacked on a pallet with a particular stacking pattern). The interface of this output to the following operations may also need to be stated, particularly if manual intervention is required at this stage.

The required throughput, or cycle time, for the system must be defined, as well as the availability required of the system, which is the percentage of time the system is expected to be operational and available to work. The actual production output required is the combination of the throughput and availability. It is not sensible to aim for a cycle time that achieves the required production output at 100% availability because this does not allow for any downtime due to maintenance, material replenishment, lack of operators or any other factor that can impact on the availability of the system.

If the system is to handle more than one product, or more than one product size, the batch sizes should be discussed. Additionally, the specification should cover the requirement for the product changeover, which could be automatic or require manual input. The need for any checking, to ensure the correct product is being processed, should also be included. In addition, it may be necessary to address any implications for the run-out to allow the system to be emptied of a product. The document should also detail any anticipated need for manual intervention and how this is to be achieved.

The customer should include some description of the controls and particularly the human machine interfaces (HMI). This may not be the detail of exactly what equipment is to be provided but rather the functionality that is expected. Such requirements include operator selectable functions, such as start, stop and access to system requests. The customer can also help the process by specifying the types of errors to be logged and how they will be logged (e.g. over what period and the level of fault diagnosis that is expected). If possible, any overall requirement for error recovery should be described. Also, the specification should include any requirement for production management information, such as parts per hour, downtime and cause. If any of this information is to be provided to an external system, this requirement must also be defined. The types and numbers of displays may be specified to provide a guide to the vendors.

Finally, the document should include any specific requirements relating to guarding, safety, and environmental issues. For example, the specification should address the construction of the guarding, such as solid panels, weld mesh or plastic, and also any preferences as to the guarding of operator access stations, such as roller shutter doors or light guards. If there are any requirements for dust or fume extraction, these also need to be included.

6.2 SCOPE OF SUPPLY

The scope of supply is also important. The previous elements of the specification basically define what is required from the automation solution, but this section details exactly what is required of the potential vendors. For example, scope of supply specifications may include the design, manufacture, assembly, testing, delivery, installation, and commissioning, as well as any training and post-commissioning support. It is also important to define the acceptance criteria. Most of the above can be covered by the term "turn-key", but that is open to interpretation, and it is better to specific exactly what is expected. This section of the specification therefore defines the expectations of the customer for each stage of the project.

6.2.1 Free Issue

If any equipment is to be free-issued by the customer, that equipment must be detailed, including the specification and supplier details. This free-issue equipment may be existing machines or new machines that the customer intends to purchase as part of the project. If these are to be integrated within the automation system, the customer needs to define the specification of these machines and how they are to be interfaced with the automation system. This interface information includes communication as well as mechanical issues, such as access for unloading and reloading. If the machines are to be delivered to the vendor during the project, the availability and timing should be detailed and also the delivery and return should be covered. If any parts are to be provided during the project, as might happen for testing, the type and availability of these and the timing should be detailed.

6.2.2 Safety

It is important to identify the safety requirements and, in particular, any customer specific requirements that should be covered. If there are any specific items that are not to be included in the scope of the vendor, such as fume extraction equipment, the specification should clarify the interface between that to be provided by the vendor and that to be provided by the customer.

6.2.3 Services

The vendor needs to be made aware of the provision of services to the automation system. In particular, the power and air, including the location and number of service drops, which are available for the system. It is also important to state who will be responsible for the quality of these services.

6.2.4 Project Management

The specification should clearly define the expectations of the customer with regard to project management. For example, the document should present the timescale within which the name and contact details of the vendor's project manager are to be provided. This may also include a requirement for the project kick off meeting to take place within a specific number of days after order placement, the objectives of the kick off meeting, the frequency and location of subsequent meetings, and the frequency and requirements for any project reports and updates to the timing plan. It is often beneficial to specify that the project manager will provide details of the project team, including any major subcontractors, at the kick off meeting and confirm the project timing plan within a short period following this meeting. There are often changes to a project once the work has commenced, and it may be worthwhile to specify how the customer and vendor will decide on any changes. This includes the level of authority required from the customer side, as well as recording and reporting requirements to ensure there are no disagreements related to variances in cost as a result of these changes.

6.2.5 Design

In this section, the specification describes the method by which the design is to be developed, including any CAD tools. The section should include the customer's need to review designs prior to manufacture, if appropriate. It may be beneficial to require the development of a functional design specification (FDS) as the first stage of the project, with appropriate customer reviews and approvals, prior to the commencement of any manufacture or software development. This may include the requirement for a failure mode effect analysis to identify risks within the project, potential solutions to mitigate these risks and the party responsible for providing these solutions. The FDS provides a detailed specification of exactly what is to be provided, the planned performance of the equipment and how it is to be operated.

The specification should explain any requirement to review the safety aspects of this FDS, particularly if specific safety personnel from the customer side are to be involved.

6.2.6 Manufacture and Assembly

The customer may wish to request certain standards, such as company standards, are applied to the manufacture of items within the scope of supply, including the software. The customer may also have specific standards for equipment or manufacturers of equipment. These must be provided, either at this stage of the specification or in the design stage. The customer may also wish to gain access, subject to reasonable notice, to view the system during manufacture and assembly, in order to view progress, and the customer should state this requirement within the specification.

6.2.7 Predelivery Tests

Prior to the shipment of the system to the customer, the vendor normally performs tests to ensure the system meets the customer requirements. These are known as the factory acceptance tests (FAT). The complexity and completeness of these tests is very dependent on the automation system to be provided and the scope of supply provided by the vendor. The tests should be designed to determine if the vendor has met the criteria required by the customer as detailed throughout the specification. The tests therefore should include an overview of the system and the component parts, as well as the overall operation and functionality.

The performance of the system should also be tested. This can range from simply the striking of a welding arc on a welding system, with the customer being responsible for the fixtures and programming, to the full operation of a system for a number of hours to prove the functionality, cycle time, and reliability. In the latter case, the customer must be prepared to make product available to the vendor both to develop and test the system, as well as the FAT. The customer should also specify who will operate the system for the tests and the period, or number of parts to be produced, to define the length of the tests. If free-issue equipment is involved within the project, the impact of this on the performance of the system must be considered. Also the consequence of any failures during the FAT, in particular how the FAT should proceed, must be specified. The specification should clearly define the actual tests to be performed, especially those that involve a measure of performance such as cycle time, reliability or quality. These are discussed further in Section 6.3.

Defining the FAT can be a challenge. It is important to undertake tests, with appropriate timescale and detail, to provide a reasonable indication that the system is performing to the specification. However, the test condition is not the real production situation, and therefore, the tests should not be so onerous that they cannot be realistically achieved.

6.2.8 Delivery

If there are any specific issues related to delivery, these should be identified in the specification. This may relate to the notice period required, the time of arrival, the offloading and positioning of equipment, the sequence of the delivery for larger pieces of equipment or the responsibility for the return of parts or equipment that has been free-issued and the removal of packing materials.

6.2.9 Installation and Commissioning

It may be necessary to specify that the vendor must conduct a site survey to confirm the location of the system and to check access and the adjacent equipment. It may also be necessary to check the floor to ensure it is appropriate both in terms of condition and strength. If the customer does not specify that these are requirements to be conducted by the vendor, the vendor will assume the customer will cover these issues.

The specification should ask the vendor to provide a layout drawing confirming the positions of the equipment within the system and also the position of the services. This information should be checked against the existing facility. The customer can provide details of the existing facility and ask the vendor to check, or the customer can check based on the information provided by the vendor. It is worthwhile to identify the approach to be taken in the specification, particularly if it is the former.

The specification should also request an installation plan at an appropriate point in the project. This should identify the number of personnel involved, the hours they are planning to work, any assistance required from the customer and also any specific requirements that may impact existing production. If there are any constraints relating to access to the site, these should be stated as well. The specification should also state that the vendor is responsible for ensuring that all its personnel, and any subcontract personnel, adhere to all contractors' site rules and health and safety rules in operation at the relevant site, including the successful conclusion of any appropriate training and the provision of all necessary safety equipment and personal protection equipment.

6.2.10 Final Testing and Buy-Off

Once the commissioning phase has been finished, there is usually a final test that, when completed successfully, provides for the buy-off by the customer and the handover of the system from the vendor to the customer. This test is often called the site acceptance test (SAT), and it follows a similar format to the FAT discussed in Section 6.2.7. The SAT is normally more detailed, with testing over a longer period or with higher volumes of product than is used in the FAT, because, at this stage, full operation under production conditions can be achieved.

The specification should describe the parameters for the tests, including the type and length of the tests. The SAT may consist of three stages, the first being an inspection of the equipment and documentation to ensure they are in line with the initial specification (URS) and any changes agreed throughout the project. The second stage is the operation of the system for a defined period to determine the cycle time, or throughput, and the quality of the product output. The final test might be over an extended period to determine the availability of the system.

As part of the definition for the tests, specification should identify the personnel who will operate the system. These may be the customer's personnel, but there may also be a role for the vendor's personnel, particularly during the availability test, because they will have better knowledge of the system and will be able to rectify any problems more efficiently. It is important to define the method by which the tests are to be measured and recorded to avoid any ambiguity. Further details are discussed in Section 6.3. Also, the actions to be taken in the event of a test failure must be defined. For example, the customer may wish the vendor to correct any problems and then restart the full SAT test or restart the relevant element of the SAT.

These tests are one of the most important elements of the specification. This is particularly true if the vendor is providing a "turnkey solution", that is, they are responsible for the complete system. The FAT is important because it can avoid the need to correct problems on the customer side, which is preferable for both customer and vendor. However, the SAT is more important because this provides the final acceptance of the system. Without a clear definition of the SAT, there can be no clear buy-off criteria and, therefore, no measurements against which the vendor can be held accountable in the event that the system does not perform as anticipated. Thus, it is most important that the customer carefully considers this element of the specification and ensures that the defined tests and measures provide a realistic assessment of the system as it should operate in production.

6.2.11 Standby

A customer may request that the vendor retain one or more engineers on site for a period after the system has entered production. This may be limited to the period of the SAT, or it may be extended beyond this period to provide assistance to the customer's engineers in the event of any problems. The requirement for standby normally depends on the complexity of the system and the experience of the customer's engineers. If standby is considered necessary, it should be specified within the URS, including both the period required and the number of shifts to be covered.

6.2.12 Training

Training is an important aspect of any automation system. There will always be a need for some training, unless the system is a repeat of a previous installation. The training may include a number of different requirements. First, there may be a need for the customer's personnel to attend robot training courses at the vendor's premises. Second, the customer will need training on the actual system, covering both operation and maintenance, including preventative maintenance, fault finding, and error recovery. This latter requirement is often covered during the period of commissioning, SAT, and standby. It may be necessary to provide training to multiple personnel, both operators and maintenance personnel, across a number of shifts. If possible, the customer should identify the numbers of personnel from each category to be trained. It is important that appropriate training is included, and the customer should at a minimum require the vendor to specify the training that is included and when and where this will take place.

6.2.13 Documentation

The customer should also specify what documentation they expect to receive, including the desired format and number of copies. This might include an operation and maintenance manual to cover the instructions for safe use of the system, error codes and fault recovery procedures, recommended preventative maintenance, and also details of recommended spare parts and consumable items. A second documentation package might include more details of the as-built system, covering items such as electrical drawings, software listings, system certifications, and the FDS. It is important that the customer defines what documentation is expected as part of the overall delivery of the system.

6.2.14 Warranty

It may be worthwhile for the customer to specify the warranty period they expect for the system, particularly if they wish to have cover for a period longer than 12 months. The point at which the warranty commences should also be specified, which may be completion of the SAT, rather than the delivery date. The specification should also state any interest in a service contract for the period of the warranty, or extended beyond this period, covering annual preventative maintenance visits or breakdown callout cover. The vendor can be requested to include these items within the proposal, either as standard or as an option to the main price.

6.2.15 Other Items

The URS should identify the main customer point of contact, including all relevant contact details. This may be multiple personnel, such as engineering and purchasing, as appropriate. If multiple personnel are listed, the responsibilities should be clearly identified to direct the vendor to the appropriate contact point. Also any preferred equipment vendors should be identified if appropriate. Relevant standards should be specified if not already covered, and any customer specific standards should be included in the package sent to the vendors. The anticipated timing for the project could also be included if the planned production start date is known and this information is not covered elsewhere.

6.3 BUY-OFF CRITERIA

The specification should clearly define the tests to be performed, particularly those that involve a measure of performance such as cycle time, reliability or quality. The definitions must include the period over which the tests are to be run and the measurements to be made. The URS should include the calculation for linking these measurements to defined performance parameters to avoid any misunderstanding or ambiguity, because this can lead to difficulties at the point of buy-off.

For example, the system could be run for a period of one shift (e.g. 8 h), operating at 100%. The output would be checked to determine the quality produced and identify any rejects. The check compares the product against the defined quality standard, which may involve dimensional checks or, if a visual quality is required, against an agreed standard, quality part. In the latter case, this is often called the "golden part". The downtime would also be recorded, including the source of the downtime. The results would be compiled, as follows:

- TO = Total output
- QR = Number of reject parts
- TT = Total time
- DV = Downtime due to Vendor supplied equipment
- DO = Downtime outside of Vendor supplied equipment

This would then allow the production rate of parts with the correct quality to be calculated:

$$\text{Production rate} = (\text{TO} - \text{QR})/\text{TT}.$$

The Availability could also be calculated:

$$\text{Availability}\,\% = 100 * ((\text{TT} - \text{DO}) - \text{DV})/(\text{TT} - \text{DO}).$$

The above calculation is intended as a simple example. It may be necessary to include other elements in the assessment and the calculation. It may also be necessary to specify other buy-off criteria to correctly assess the performance of the system. The length of the system's operating time for the buy-off is very dependent on the type of application and the production rate. It should provide a reasonable test of the system, covering both capability and reliability.

6.4 COVERING LETTER

The URS could be accompanied by a covering letter that can address items not covered within the specification. This includes specific items such as the timing of the proposal stage, including any arrangements for visits to the customer site, the desire for any vendor presentations and potentially when the order is expected to be placed. The cover letter should also state the deadline when the quotes are to be received. In addition, it should outline the customer expectations regarding the responses from the potential vendors, or what the customer expects to see within the proposal. This may include:

- Timing plan (from order placement to full production) including significant milestones
- Outline of the system design, including a layout drawing and main dimensions

- Details of the main items of equipment
- Description of the modes of operation, such as start up, production, maintenance, and servicing
- System throughput and availability
- Any limitations of the system
- Details of any proposed subcontractors and the equipment/services they are to provide
- Any specific floor and foundation requirements
- Services requirements (including location on the above layout)
- Price
- Acceptance of the customer terms and conditions, which should be included in the package
- Any exclusions

The anticipated delivery date or start of production date should also be mentioned if it is not covered in the URS. It would also be appropriate to include the payment terms for the project if they are not covered elsewhere. Additionally, the cover letter can ask the vendors to provide details of any reference sites performing similar applications.

6.5 SUMMARY

The URS is an important document because it defines what the customer wishes to purchase. The intention should be to provide all the information needed by the vendors in order to develop a proposal to meet the customer's requirements. The URS therefore provides a standard against which the offers from the different vendors can be assessed, and it provides greater reassurance that the vendors are all quoting against the same requirements.

The URS not only assists the customer, but it also assists the vendors because they are provided with a clear description of the requirements of the project. The more experienced vendors will welcome this document because it does ensure a consistent basis for quotations, thus removing the opportunity for some vendors to provide low prices based on assumptions regarding the scope of supply and then potentially winning the project without the same content as other suppliers (see Chapter 8).

The URS also provides protection to the customer in the event that disputes regarding the scope of supply arise later. The clear definition of the buy-off criteria is a key element of the URS because this ensures that there are measurable parameters to be used in testing the system prior to acceptance. If these are defined prior to quotation, the vendors are aware of what will be required, and they can therefore include the costs necessary to cover this work.

For these reasons, the customer should develop a URS for all projects. This may be difficult for some customers who have less experience with this approach or the application of robots and automation. Appendix provides an example of a typical URS and cover letter. If new to the process, the customer might benefit from bringing in an external resource to assist with the development of the URS. Although there will be costs associated with this external resource, the reduction of risk and the related financial consequences often outweigh this investment.

CHAPTER 7

Financial Justification

Chapter Contents

Abstract

This chapter discusses the requirements for the financial justification of a proposed automation solution. The specific savings that may be available from each of the 10 key benefits of robots are reviewed together with some examples of how these can be expressed in financial terms. An approach towards developing a financial justification is discussed, including both savings and also other project costs that may be missed. The importance of providing the appropriate budget for a project is also highlighted.

Keywords: Payback, Budget, Justification, Labour turnover, Scrap, Rework, Return on Investment, ROI

Most projects will require the development of a financial justification to provide the budget against which the project can be purchased. The justification normally provides a comparison between the current costs of performing an operation or process, often using manual labour, and the savings that can be achieved by the introduction of automation. These savings are then compared to the anticipated cost of the automation solution to determine the point at which the investment starts to generate a positive financial return, often called the Return on Investment, or the ROI.

Most companies will apply a standard period, or payback period, over which investments are considered to be worthwhile. If the expenditure is covered within this payback period, then the project is considered financially viable and, to that end, worth considering. This is a major hurdle that must be addressed before a project can proceed. The decision to accept the project as viable often involves senior management, including financial personnel. It is, therefore, important that the benefits of the automation project be expressed in terms and language that can be understood by these people. The justification needs to be expressed in financial terms using criteria that are recognised and accepted by senior management. The development of these justifications can be a challenge for the project engineers because they often lack the knowledge and training to describe the benefits of the project in the appropriate way. However, the process is logical and it is not difficult to express the project using the appropriate terminology.

The main challenge is often the identification of the likely cost savings. These are not just the most obvious, typically the saving of labour. Based on labour savings alone it can be difficult to achieve the target payback period. This is particularly true in countries such as the UK and the USA where relatively short payback periods, such as 12 or 18 months, are the norm. By careful study it is often possible to identify additional savings that, by reducing the payback period, can make projects viable. This chapter reviews the process and, in particular, identifies examples of less obvious cost savings and explains how these might be derived.

It is also worth noting there are situations where the use of automation is not compared with alternative methods of manufacture. These are within companies where the use of robots is already fully proven, such as the automotive or automotive component sectors. In these cases, it is often a given that robots will be used for a specific project. The question is more about how many robots are required or how the systems will be designed. The justification for the purchase of the robots is not normally required because this approach is the norm. However, a justification may be required if a new automated process is being considered; in which case, the comparison is between the costs of the alternative automated solutions and the savings that are predicted for the new solution. The process and investigation of potential cost savings would follow the same approach as that discussed below.

7.1 BENEFITS OF ROBOTS

The 10 key benefits (International Federation of Robotics, 2005) that can be obtained by the application of robots were introduced in Chapter 2. These are as follows:

1. Reduce operating costs
2. Improve product quality and consistency
3. Improve quality of work for employees
4. Increase production output
5. Increase product manufacturing flexibility
6. Reduce material waste and increase yield
7. Comply with safety rules and improve workplace health and safety
8. Reduce labour turnover and difficulty of recruiting workers
9. Reduce capital costs
10. Save space in high-value manufacturing areas

Each of these is reviewed below in terms of how they might contribute to the financial justification for a project. It should be stressed that these are potential benefits and may not be applicable in all cases.

7.1.1 Reduce Operating Costs

There are a number of operating costs that can be reduced by the introduction of robot automation. The most obvious being the labour cost previously incurred by the manual performance of that operation. Associated with this labour cost is the overhead that can be attributed to the labour; that is, the cost of employing the staff. This cost includes a proportion of the human resources function, payroll and supervision, training, and health and safety, as well as other costs such as the canteen and personnel protective equipment (PPE). If the number of personnel being replaced is small, these costs may not be significant. Additionally, if the labour is being redeployed elsewhere within the operation there may not be any actual reduction. However, even if the labour is redeployed there would have been additional headcount required if the automation was not introduced, so a saving is produced and the associated overhead should be included as part of the justification.

It may also be possible to reduce energy consumption. The energy per unit of output is optimised by the consistency of the robot and the reduction in scrap or rework. Additionally, robots do not require the same heating (or cooling) and lighting necessary for manual labour. Therefore, if the particular area of the factory can be fully automated, then it is possible to reduce energy input required to maintain the working environment.

7.1.2 Improve Product Quality and Consistency

The consistency of the robot automation will reduce defects as well as providing a consistent production rate. The reduction of these variables provides savings that should be measurable. The known throughput can lead to reductions in overtime or other administrative costs related to dealing with a variable production output. The reduction in defects will reduce the scrap rate as well as any rework requirement. This leads to savings in the labour, tools, and facilities required to perform the rework. The reduction in the scrap rate not only leads to savings as a result of the reduced production time required to produce the necessary quality output but also the cost of dealing with the scrap.

7.1.3 Improve Quality of Work for Employees

By improving the quality of work, the workforce is likely to have increased motivation and, consequently, productivity and quality will improve. The introduction of new manufacturing facilities also increases the confidence of the workforce because it shows the company has belief in the production facility and is willing to invest in new equipment. The investment provides opportunities for new jobs, such as maintenance and programming, which again has a positive impact on attitudes. Overall, these positive impacts provide increased workforce motivation but this is very difficult to quantify and, therefore, show as a financial benefit.

There may also be a reduction in labour turnover as the unpleasant jobs are those most likely to have high labour turnover. This reduces the costs of hiring new staff (see Section 7.1.8) and the human resources costs related to training and introducing new staff.

7.1.4 Increase Production Output

As mentioned above, the consistency of robots ensures a regular and potentially increased production output. This can increase the output of other machines, providing more efficient utilisation of those resources. It is also possible to operate machines unattended, which means they can continue to produce during operator breaks or can be set up to run at the end of a shift, overnight, or at weekends.

The value of this increased output can be determined. In some cases it may mean there is no need to purchase additional machines leading to significant savings, not only in capital and running costs but also in space and energy.

7.1.5 Increase Product Manufacturing Flexibility

The flexibility of robots can reduce changeover time between different products, which provides cost benefits in terms of reduced downtime. The reduction in changeover times can also allow the production of smaller batches, reducing the costs of storage and work in progress.

7.1.6 Reduce Material Waste and Increase Yield

As discussed in Section 7.1.2, reducing defects leads to cost savings in terms of the handling of those defects. It also leads to savings from a reduction in the costs of the materials fed into the system because these are utilised more effectively. This reduces the work in progress through all processes prior to the automation, including goods in and all the operations and activities associated with handling those materials including, potentially, stock control, and purchasing.

In addition to increasing the yield from the input materials, by reducing the production of defects, the automation can also work to tighter tolerance bands and, as a result, only apply or use exactly what is required. In some industries this can be a significant cost saving. For example, in the food industry minimum pack weights must be achieved and the use of automation allows the weight to be more tightly controlled reducing the amount given away to guarantee the minimum pack weight. This also applies to the use of consumables, which are used more efficiently by automation and, then, not only is the cost of the consumables reduced but also the cost of replenishing and storing those consumables.

7.1.7 Comply with Safety Rules and Improve Workplace Health and Safety

In addition to the improved production output and quality that can be achieved by automating dirty, dangerous, and demanding tasks (which would be covered by Section 7.1.2) other savings can be achieved. First, the requirement for PPE is reduced, which can also reduce the indirect costs associated with the PPE, such as purchasing and storage. There is also a potential reduction in employer insurance costs because the risk of claims due to injury is reduced.

There may also be a reduction in labour turnover as the unpleasant jobs are those most likely to have high labour turnover. This reduces the costs of hiring new staff (see Section 7.1.8) and the human resources costs related to training and introducing new staff.

7.1.8 Reduce Labour Turnover and Difficulty of Recruiting Workers

The most repetitive, dirty, or demanding jobs are those that normally have the highest labour turnover. The opportunities that automation provides generates new roles that are more challenging, less repetitive, require higher levels of skill, and also lead to higher levels of pay. The investment by the company also demonstrates confidence from the management and, consequently, generates a positive attitude within the workforce. All these factors, particularly the removal of the most arduous or repetitive jobs, leads to a reduction in labour turnover.

By improving staff retention, the costs associated with hiring new workers are reduced. This includes not only the direct costs associated with the hiring process but also training costs and the cost of workers' lower productivity as they come up to speed on the job.

7.1.9 Reduce Capital Costs

As mentioned in Section 7.1.5, the flexibility provided by robots can enable smaller batch sizes, which reduces the cost of work in progress and inventory. Robots can provide the opportunity to run other machinery more efficiently or perhaps over extended hours, which may mean it is unnecessary to purchase additional machines (see Section 7.1.4).

The ease with which a robot system can be reconfigured can also provide longer-term capital cost benefits. If a product design change is required or a new product is introduced, a robot system can be reused. There will be some cost associated with the reconfiguration but this would be lower than the cost required for a completely new machine.

7.1.10 Save Space in High-Value Manufacturing Areas

An automation system can often be much more compact than an equivalent manual system. The more efficient utilisation of the equipment involved and the greater production throughput can also result in a reduction in the floor space required.

This space saving has a value that can be included in the financial justification. In some cases, the use of automation could remove the need for a building expansion, which provides a very significant financial benefit.

7.2 QUICK FINANCIAL ANALYSIS

It is often worthwhile performing a quick check of the financial benefits to determine if a particular project has any chance of being approved by senior management. This quick check would determine if the financial payback is likely to achieve the internal payback requirement; that is, the number of months required for the financial return to cover the cost of the investment.

The first step is to determine the budget cost for the automation system. This will require the development of an initial concept (see Chapter 5) and the determination of the anticipated cost of this solution. It may be worthwhile involving suppliers at this stage for two reasons. They may be able to assist in the development of the concept as they may have ideas, based on their experience and knowledge of the capabilities of their products, which provide a better solution. They may also be able to provide a better indication of the likely cost of the system; again, based on their experience and also their knowledge of what is required to deliver the solution proposed.

Once the concept is defined it will be possible to identify the direct labour saving that results from the implementation of that solution. Please note that if the system is to operate over multiple shifts it is the total labour saving – that is, the sum of labour savings per shift – that should be considered. The total labour saving can be calculated from:

$$\text{Total annual labour hours saved} \times \text{Cost per labour hour}$$

or:

$$\text{Total labour saving per production unit} \times \text{Cost per labour hour} \times \text{Annual volume}$$

Either of the above will provide the cost saving resulting from the labour reduction.

If there are other very obvious savings it would be worthwhile introducing these as well. If the introduction of automation means increased throughput can be achieved, to meet increased production demands, there may be additional saving from the extra labour that is not required to meet this increased production or the savings that result from other machines and facilities that will not be required. The total value of these savings can be calculated on a monthly or yearly basis.

The payback is then determined by dividing the estimated cost by the total saving to provide a payback in months or years (dependent on whether the savings was calculated on a per month or per year basis). This payback can then be compared to the criteria specified by the company to determine how close to the target payback the project would be.

Having determined this budget payback period, it is then worthwhile to consider the factors discussed next.

7.2.1 How Conservative Is the Calculation?

This decision is based on the knowledge of what is included in the budget costing and what additional expenditure may be required, such as internal training and the ramp-up time from the completion of the installation to full production (see Section 8.1). Also, there may be known additional savings that can be quantified, given further work, which would shorten the payback. The assessment of these issues will provide an understanding as to the likelihood of the payback period either reducing or increasing given further analysis.

7.2.2 What Is the Technical Risk?

If the concept uses proven equipment and solutions that provide a high level of confidence that the project will be trouble free, then it is likely the anticipated payback will be achieved. If the project requires unproven solutions or is a significant step forward in technology there should be some allowance for the associated risk and the likely costs that could result, thereby increasing the payback period.

7.2.3 Is the Solution Flexible?

The concept may provide the opportunity to include other products and, therefore, improve the utilisation and payback, or it may be dedicated to a specific product, limiting the payback to the savings achieved on this product.

7.2.4 What Is the Driver for the Investment?

The requirement for the automation solution may be driven by customer requests for your product. This may be due to quality requirements, increased output, or even products that can only be produced using automation. In this case, the investment decision will need to consider the implications of the retention of this business and also the impact if this business is lost, potentially including other business that is conducted with that customer.

7.2.5 Is the Solution Future Proofed?

If the concept contains a significant element of equipment dedicated to the product for which it is designed, the flexibility of accommodating changes in product design or changes in production volumes, by introducing other products, will be limited. The justification can, therefore, only be based on that specific product, and there may be a future risk if that job is lost or the customer changes the product design.

If the concept has a high degree of flexibility and the cost of introducing new products or product changes is small, the system can be reconfigured for limited additional cost. The automation solution is thereby future proofed and could potentially be justified over a longer period.

7.2.6 Competitive Position?

It may be known that competitors have or are introducing automation and, as a result, may gain competitive advantage. This, in itself, should not be an overriding factor because the competitors may be making a mistake but this factor should be considered, particularly if the project is implementing proven technology and, therefore, is not seen as a risk.

7.2.7 Company Attitude to Automation?

In some companies there is a very positive attitude towards automation, and senior management believes in the benefits and understands the need for automation. In these cases they are likely to be receptive to a proposal even if the justification demonstrates a payback that is close to or slightly over the defined cutoff. The senior management is also likely to accept some of the less direct savings that can be identified as a benefit from the automation (see Section 7.3). In most companies where successful automation projects have been executed and the savings have been demonstrated, management is often more supportive because it has seen the positive results that can be achieved.

In companies where management is less aware of the benefits of automation, the justification is much harder and management is less likely to support projects that require less tangible savings to achieve the justification or the payback period is relatively long.

7.2.8 Project - Go or No Go?

Having performed the quick payback calculation and considered the other issues identified above it should be possible to determine whether it is worthwhile developing the project and the justification further. If the driver is an end-customer requirement or it is known that competitors are already introducing automation, the decision may be simple and the payback criteria may not be as important.

If the project is not driven by external issues, then the payback is important and the decision needs to be made as to whether further work could improve the anticipated payback period, if that is required. If it is determined that the project should continue to the next stage, it is worthwhile reviewing the sources of additional savings because this can assist in ensuring the appropriate budget is allocated (see Section 7.5).

7.3 IDENTIFYING COST SAVINGS

As discussed above, in addition to direct labour savings there are many different cost savings that could result from the implementation of the automation solution. Examples include the following:

- Fixing product quality and inconsistency
- Improved safety
- Increased manufacturing flexibility
- Improved operations reliability
- Improved regulatory compliance
- Increased product yields
- Increased productivity
- Reduced manufacturing costs
- Reduced scrap or rework
- Reduced floor space

To determine which of these, or others, may be applicable requires a study of the current situation and the likely benefit of the particular automation solution resulting in a comparison of the two. It may be necessary to involve finance, human resources, and other departments to identify some of the costs because these would fall outside the data normally accessible to manufacturing. However, this can be an enlightening activity because some of these costs may not have been directly linked to specific production operations, from which they originate, and may also be higher than anticipated.

Once the list of likely benefits has been identified, the next step is to determine the cost savings that could be obtained. Some examples of how the savings might be calculated are provided below as a guide.

7.3.1 Quality Cost Savings

If rework is currently undertaken and the consistency of the automation is expected to remove or reduce the need for rework the saving can be calculated from:

$$\text{Total annual rework labour hours saved} \times \text{Cost per labour hour}$$

or:

$$\text{Current rework costs} \times \text{Rework reduction }\%.$$

If there is currently an amount of scrap produced and this is to be reduced the saving can be calculated:

$$\text{Yield improvement }\% \times \text{Annual production volume} \times \text{Unit cost}$$

or:

$$\text{Annual reduction in scrap produced} \times \text{Unit cost}.$$

Likewise, there may be an anticipated reduction in warranty costs due to the improved and consistent quality. This could be determined from:

$$\text{Total number of warranty failures} \times \text{Reduction in failures }\% \times \text{Cost to rectify}$$

or:

$$\text{Current warranty costs} \times \text{Reduction in failures }\%.$$

Many companies capture warranty costs as a percentage of sales and therefore this cost can be identified.

7.3.2 Reduced Labour Turnover and Absenteeism

If the operation to be automated suffers from an unusually high labour turnover rate, then the costs associated with replacing the labour can form an element of the justification. These would be determined as follows:

$$\text{Average hiring cost, per person} \times \text{Number of positions saved}$$

Plus:

$$\text{Average time to train} \times \text{Labour cost per hour.}$$

Likewise, if the operation has a greater level of sickness or absenteeism than is normal for the rest of the production operation, the additional cost of labour to provide cover and the training required for that labour can be included in the justification.

7.3.3 Health and Safety

In addition to the costs of the PPE, which can be quickly calculated by the PPE cost per person per year multiplied by the reduction in the number of jobs, there may also be a reduction in claims due to injury or illness. The latter example can only be used if there have been claims made by workers who are directly working on the operation to be automated, but if these claims are made their inclusion in the justification can significantly improve the payback period.

7.3.4 Floor Space Savings

The company may have an annual floor space cost, which could include maintenance, heating, and lighting. If this is known or can be obtained and if the automation solution reduces the floor space required for an operation, the saving in space can be converted into a financial saving that can then be inputted into the payback calculation.

7.3.5 Other Savings

The calculation of other savings would be done on a similar basis to those illustrated above. For example, if the consumption of consumables is expected to be reduced, the cost saving would be the percentage reduction multiplied by the annual consumable cost. If the yield from the input material is to be increased through the greater control and consistency of the automation, the saving would be the percentage increase in yield multiplied by the cost of the input material.

The above discussions are based on direct labour cost. This would not necessarily include the administration, training, human resources, and other costs associated with the labour. These additional costs would be included in the overhead costs or the burdened labour cost. It is certainly justifiable to use these higher costs in any justification. If the total labour cost per person, including this overhead, can be obtained and used in the examples above as well as the labour-saving calculations, it will assist the financial justification.

7.4 DEVELOPING THE JUSTIFICATION

In many cases potential end users only consider the cost benefit of the direct labour saved as a result of the automation, but often the financial justification can be improved by also assessing the other benefits. In some cases, significant savings from another benefit may outweigh the saving in labour cost, resulting in the achievement of acceptable payback periods that otherwise might not have been achieved.

The justification should be based on all the relevant factors; that is, all the potential cost savings. These should be identified in turn with the rationale provided detailing why the automation will achieve a specific benefit and the anticipated annual cost saving resulting from that benefit. It is beneficial to start with the direct labour savings, followed by other tangible savings, with the less tangible items at the end. In cases where the benefit and, therefore, cost saving is not guaranteed, it is better to provide a range for the cost saving to indicate what may be achieved along with the rationale explaining the limits of the range. This approach provides credibility to the justification because the uncertainty will be understood. Having itemised all the annual cost savings the total can be determined, showing both the minimum and maximum anticipated saving based on the ranges for some of the items.

The budget cost for the automation system will be included based on internal estimates or budget quotations provided by suppliers. It is important that the costing covers all the items required to deliver the project and so a number of other items will be added to cover costs associated with the internal resources required for the project. This might be travel expenses to visit the supplier, training costs, and possibly an external consultant to assist with the execution of the project. It is also important to consider the potential for disruption to production during the installation and commissioning phase, and the likelihood of incurring additional costs as a result of this disruption. There is likely to be the need to provide parts for trials, programme development, and preproduction runs. These parts are often scrapped afterward and, therefore, the cost of these parts should be included. It may be worthwhile including a contingency, based on the perceived risk involved in the project (see Section 7.2.2). All these cost items should be identified and the rationale explaining the cost included. In particular, it is worthwhile identifying and explaining any contingency as this demonstrates a conservative approach.

It may also be worthwhile explaining any factors that have not been covered in the financial case. These may be items such as the market perception of the investment, the potential to impress future and existing customers, the opportunity for publicity, the motivation of the workforce, and other benefits that would be impossible to quantify financially.

Finally, the payback period can be calculated based on the total cost and the total savings. If ranges of numbers are included in the previous calculations, the result will also be a range. If the company has a specific cutoff for payback periods, there is little value in providing a payback outside of the allowable range as the project will not be approved.

The process may be iterative in that the first pass may not meet the payback criteria but with further work, either on the automation concept and costs or the identification and valuation of the anticipated savings, it may be possible to shorten the payback period. However, it is not worthwhile adjusting the numbers to achieve the desired payback period if this is not realistic (see Section 7.5).

The objective is to achieve a payback within the desired company objectives based on realistic project costs, realistic benefits, and cost savings for activities or items that are recognised by the company as valid to include in a cost justification. If this can be achieved, the project has a chance of being approved. However, it should be recognised that senior management may be considering a number of different projects from different areas of the company. The total funds available for investment may be constrained and so this project may be competing with others for the available financial resource. Accordingly, the calculation of the anticipated payback must not be too cautious because the project may not proceed as other projects with better anticipated paybacks may be preferred.

7.5 NEED FOR APPROPRIATE BUDGETS

As a final comment regarding the development of cost justifications, it is imperative that the budget proposed provides enough resources to execute the project successfully. The wrong approach is to use the payback period as the main driver, reducing the budget to a level where the necessary payback can be achieved.

Under this scenario, it is likely that the budget does not provide the resources required to execute the project successfully. It may be that internal resources cannot be utilised or the lowest cost supplier has to be selected (see Section 8.2) to meet the constrained budget. As a result, there are likely to be problems at some point during the project that will result in delays or unforeseen costs. Neither of these outcomes is beneficial, and it is unlikely the expectations of senior management will be achieved. The final result will either be an automation system that does not perform as planned or one that costs more than planned. The anticipated payback is, therefore, unlikely to be achieved and in the worst case there may be longer-term reliability or performance issues. If this type of project does occur, it not only affects this project but also provides a barrier to future investment in automation by the company.

Hence, if the desired payback cannot be achieved, based on the anticipated project costs, it is better to reconsider the benefits and anticipated savings to ensure these have been fully identified and assessed.

The overall objective of the justification process is to identify reasonable cost savings that, by showing a payback period within the criteria set by the company, justify the budget required to execute the project successfully, and thereby deliver the anticipated cost savings.

CHAPTER 8

Successful Implementation

Chapter Contents

Abstract

This chapter identifies the main stages in successfully implementing an automation project after the specification and justification processes are complete. We review the selection of vendors, as well as the planning and execution of the project. To this end, we discuss the benefits of involving staff and vendors in the various stages of a project. The chapter also includes a review of common problems and how these might be avoided.

Keywords: Project planning, Vendor selection, Failure mode effects analysis, Installation, Commissioning, Documentation, Staff involvement

The successful implementation of an automation project requires a project manager who possesses the skills and techniques needed for any capital investment project. An automation project does require a team with automation knowledge and expertise, but the project manager does not necessarily have to have this skill set. Instead, with the necessary technical support, the project manager can provide planning, budget control, and more general project management functions.

The development of an automation project generally follows an iterative process. As described in Chapter 5, this process commences with the definition of the operations to be automated and the development of the initial concepts. These are then evaluated to determine both risk and cost, with the concept being refined until the optimum solution is derived. Once the concept has been finalised, the budget can be determined, the financial justification developed (see Chapter 7), and the project submitted for approval. Once the project is approved, a detailed specification can be created (see Chapter 6). This specification may be based on work already conducted, but it is important to commit the concept and requirements to paper to provide an expectation against which potential vendors are able to quote.

The project is now real and enters the implementation phase. The main steps in the implementation are vendor selection, system build and buy-off, installation, commissioning, and ongoing production. These are detailed below, together with guidance on project planning, staff and vendor involvement, and the avoidance of problems that can occur during the implementation.

8.1 PROJECT PLANNING

The first important step is to develop a project plan. The specification may have defined outline timing, indicating the planned order date and the required date for start of production (SOP). However, it is important to develop a much more detailed timing plan covering all aspects of the implementation, including both external vendors and internal resources.

The first important issue is to review the delivery dates proposed by the bidding vendors to determine if these are within the initial timing provided in the specification. If they are all within the requested period, then it is safe to proceed using the original timing. If one or more are outside the period, discussions should be held with the relevant vendors to identify the reasons behind their proposed timescale. The vendors may have resource issues due to existing workload, or they may have items on long lead times within their proposals. Yet, they may also have provided a more realistic delivery schedule than those who are claiming to be able to achieve the requested date. Those vendors who claim to be able to achieve the required date should also be questioned to ensure the issues identified by the longer lead-time vendors are not applicable to the shorter lead-time vendors. If the investigation determines that the original date is a risk, then it may be better to delay the installation. There should also be some contingency built into the timing plan, partly for unforeseen items but also to accommodate project stages that might cause delays. This might include the time required for design approval (see below). It is better to ensure that the expectations of the senior management regarding delivery and SOP are realistic rather than missing the planned dates in the future.

Having finalised the delivery date, the project manager should take note of the lead times quoted by the vendors in respect to the anticipated order date. Internal knowledge of the vendor selection requirements and order sign-off process, in particular the likely time required, should be considered to ensure an order can be realistically placed within the timescale requested by the vendors. This again may cause the project manager to reconsider the delivery date.

The overall project timing plan can then be developed commencing from the current position. This should therefore include time to assess the quotations and expertise of the potential vendors. If necessary, the project manager should allow time to visit existing customers of those vendors, conducting detailed discussions with a number of vendors to ensure that the specification is fully met by their quotations. A date for the selection of the vendor can therefore be identified. The vendor selection is followed by a period to allow for the generation and sign-off of the order. This period should include time for discussions over terms and conditions and the finalisation of the scope of supply and pricing. The date for order placement can then be specified.

From here, the project manager can use the vendor's lead time to identify the time required before delivery. The lead time may include a number of phases such as design, manufacture, system assembly, and proving, prior to the factory acceptance tests (FAT) conducted on the vendor's premises. It may be worthwhile to identify these stages, particularly if input is required by the customer at any stage. For example, this might include design approval and parts for programme development and the performance of the FAT. In defining this element of the timing plan, the project manager should include any delays that may be caused by the customer (e.g. the time required to provide design approval).

There is normally a need for a number of project meetings during any project. The number and frequency of the meetings will be dependent on the size and complexity of the project. As a guide, it is often worthwhile to plan for monthly meetings at the start, even if these scheduled meetings are required to change later, due to the specific circumstances at the time. The dates of these meetings can be included on the plan. This stage of the project is concluded by the FAT, after which delivery takes place.

Following delivery, the project manager should identify a period for installation and commissioning. This is followed by the site acceptance test (SAT). Production start then follows the SAT. The team might also benefit from allowing a further period for production ramp-up. This is to provide a period during which the workforce is able to gain experience of the new automation solution with the objective of achieving full production at the end of this period. If the automation system is simple, the ramp-up period may be less than one day, but if it is a complex system that requires significant learning by the operators and maintenance staff, a few weeks may be appropriate.

The team needs customer resources and products during these stages, and this need should be identified on the plan to ensure other departments can be made aware of what will be required and when it will be needed. Training may also be required, possibly including operator training, robot programming training, and maintenance training. The timing of each of these packages and the manpower to be allocated should be included on the plan to ensure the necessary resources can be made available as required.

The basic timing plan developed above forms the initial plan for the project. This includes the major milestones, such as order placement, the project kick-off meeting, design approval, FAT, delivery, SAT and the SOP. If at all possible, the plan should also include a contingency in the event of any unforeseen delays. The challenge is to ensure the SOP is both realistic and meets the needs of the business.

Once the vendor has been given their order, the project manager often requests a detailed timing plan from the vendor at the project kick-off meeting. This more detailed plan can then be compared with the initial plan, and the team can make any adjustments needed to provide a final timing plan for the project. This final plan should then be used to monitor the progress of the project and to highlight any changes to timing, giving the team the ability to assess the implications of those changes.

8.2 VENDOR SELECTION

The initial key step in the selection of vendors is to provide a specification (see Chapter 6) against which vendors must quote. This also provides a common basis against which the various quotations can be assessed, and it can be used to ensure that all the vendors are quoting to deliver equipment and services intended to achieve the same objective. The actual equipment and services may be different between each vendor because they may not all be quoting the same concept.

Normally, the project manager approaches a number of vendors for quotations. The optimum number is probably three, because this number ensures that the workload required to discuss the project and assess the responses of those vendors is not too time consuming. The project team must perform some initial work to identify the vendors to whom the customer will send the request for quotation, because the invited firms must have the necessary expertise and resources for the project. There is little value in investing time and effort in working with a potential vendor, if that firm either does not quote or falls short in the vendor assessment at an early stage (see below). The vendors also invest time and resources into the development of their proposals, and therefore, only vendors who have a reasonable chance of success should be requested to quote.

Having received the quotations, the project team must then perform a comparison of the proposals and the experience and expertise of the vendors. A first step is to review the proposals to ensure they have covered all the elements of the specification. The vendors may have listed exclusions or exceptions to the specification, and further investigation of these may be appropriate in order to determine the reasons for the exclusion and also implications for the customer. A careful review of the quotations may also lead to the identification of items that require clarification. One or more vendors may have identified issues or points which are not apparent with the other vendors. One or more vendors may have also proposed an unusual concept that requires more detailed investigation to determine if it is appropriate.

It is therefore worthwhile to hold a review meeting with each of the vendors to work through their proposals step by step to ensure the entire specification is covered or justifiable reasons and appropriate alternatives are in place for those items not covered. It is not ethical to share the ideas of one vendor with the others, and therefore, the discussions with each vendor should be treated as confidential.

Having held these meetings, the project team should understand how each of the vendors stands versus the specification, and a valid comparison can then be made. To assist with this assessment, the team may develop a table of key items against which the vendors can be scored. These key items can also be weighted to ensure that their relative importance is reflected in the assessment. An example of such an assessment is provided in Table 8.1.

An equally important element of the vendor selection is an assessment of the vendors' experience and expertise. If a vendor has experience of a very similar installation or has very relevant expertise, that firm has knowledge that will both assist with the project and also reduce the risk. The vendor will also be quoting from a more knowledgeable position than vendors without the experience, and therefore, the experienced firm's quotation is likely to include any lessons learnt from previous projects. This experience or expertise should also be included in the overall assessment discussed above, and it may be given the same importance as the price to ensure the risk of choosing a less experienced vendor is correctly factored into the assessment.

Table 8.1 Proposal and Vendor Assessment

Criteria	Scored Out of	Weighting	Importance (%)
Solution meets specification	5	×5	25
Technical complexity of solution	5	×3	15
Relevant experience of vendor	5	×3	15
Price and payment terms	5	×4	20
Project timing	5	×2	10
Service and technical support, including warranty	5	×2	10
Training	5	×1	5
Total			100

The ability of the vendor to provide service support may also be important. For example, if the system is going to be critical to the production operations of the customer and will be operating over three shifts, the ability to provide 24-hour service support with guaranteed response times may be a key issue. This type of capability is often only available via the larger vendors, because the smaller vendors do not have the necessary resources. The self-sufficiency of the customer and the need for service support from the vendor could therefore be key requirements.

Finally, the size and financial stability of the potential vendors may be considered. These factors become more important for larger projects for which the ability of the vendor to fund the cash flow through a project can be an issue. The longer term viability of the vendors is also important because the customer wants the vendor to be available to resolve future problems with the system and to possibly provide for upgrades or reconfiguration if required. Therefore, the project team must assess the longer-term prospects for any vendor being considered for a project.

Once the assessment has been completed, the results may indicate an obvious choice for the project. This could be one vendor that meets all the criteria and also achieves the best score on the comparison. If there is more than one remaining candidate, further discussions with the most appropriate vendors may be required to finalise the choice. These discussions may involve negotiations on price. If the customer's purchasing department is involved, price negotiation is normal. It is better not to push too hard to achieve price reductions because this could impact the project. Every vendor aims to achieve a profit, and it is important for the customer that the vendors do achieve this profit. Without a profit, many vendors try to adjust the scope of supply or the quality of the equipment selected to ensure they can recover their planned profit position. If the project produces a loss for the vendor, this could also affect their longer-term viability. Therefore, price negotiations should be conducted to ensure that the best deal for the customer, while also taking into account the implications of any price reductions that are achieved.

It is often a false economy to select the vendor with the lowest price, particularly if the above assessment is not undertaken. This vendor could have made errors within their cost estimates or have underestimated the resources required to execute the project, which would then lead to a lower sell price. As the project is executed, the vendor will realise the error, and, given that no companies are in business to make losses, the vendor will then try to recover the situation, which can only lead to problems within the project. If the lowest price vendor meets all the necessary criteria, but there are still concerns, it is valid to add a contingency to cover the risks. If they are still the best choice, then the order can be placed with them, but the addition of this contingency may then make other vendors more attractive.

An approach taken by some customers is to request five quotations. The lowest price quote is then rejected because the vendor is likely to have made an error or underestimated the scope. The highest price quote is also rejected because the vendor either does not want the business, possibly because they are already busy, or they may have provided an inappropriate solution. The discussions then proceed with the three vendors who provided pricing in the middle of the range.

In some cases, a customer may only ask one vendor to provide a quotation. This is due to the previous experience of that vendor and satisfaction with their previous work. This single-sourcing approach often saves time for both the customer and the vendor. If the current project is similar to previous projects, these can be used as a guide when assessing the price that is being quoted. However, if the project is a new application, it can be difficult to assess whether that vendor is providing value for money, because there is little information on which to base the assessment. Single sourcing can lead to the development of a long and mutually beneficial relationship between vendor and customer. The vendor is able to enter into more detailed conversations early in the process, and they actually form part of the team that develops the concept and the project. However, it is best to check the market occasionally to ensure the customer is receiving value for money.

Having selected a vendor, the customer can place the order. This should reference the important documents related to the project, including the specification (URS), the vendor's proposal and any other documents passed between the parties that detail clarifications to these documents. The required delivery date should be clearly stated, as should the terms and conditions against which the order is placed.

8.3 SYSTEM BUILD AND BUY-OFF

Having placed the order, the customer's project manager first holds a kick-off meeting between the customer's project team and the vendor's project team. This meeting is normally organised between the respective project managers. The purpose of the kick-off meeting is to review the entire project, covering the operations to be performed within the system, the products to be processed, the scope of supply, and the project timing, in order to ensure there are no misunderstandings between the vendor and the customer. It is worth noting that the vendor's project team is often composed of different personnel than those involved in the sale, and it is in the customer's interest to ensure all aspects of the project are fully understood and agreed upon.

This meeting should be recorded in minutes agreed upon by both parties to avoid any misunderstanding that may become apparent later. In addition, the vendor should also produce a detailed timing plan very shortly after this meeting to ensure any variance to the planned timing is apparent to the customer. This timing plan should include all the milestones (see Section 8.1), and it becomes the document against which the progress of the project is assessed.

In most projects a design phase follows. This phase covers the detail design of all the elements of the system and the final design of the overall system. Simulation is often a valuable means for providing confidence that the tools, such as welding torches or grippers, can access the parts as intended and also that the system performs to the planned cycle time. The simulation can build on work already performed in the concept development phase (see Section 5.4). Once the designs are completed, they should be submitted to the customer for approval. The customer should check the designs to ensure that they are appropriate for their use and meet the relevant standards. The customer should also verify that the design of the system will fit within the factory space allocated and the operation of the cell, including access for maintenance, is suitable for their production environment, providing a workable solution.

It may be appropriate to conduct various assessments of the design at this stage, particularly on more complex systems or applications. These assessments could include a failure mode effects analysis (FMEA) on both the process and the equipment. The purpose of these studies is to identify potential causes of failure, as well as the likely frequency and seriousness of that failure. The studies should be conducted jointly between the vendor and customer, and they should include not only aspects of the system itself but also the impact of input part variability as well as other issues external to the system that could affect system performance. The output from the studies is assessed, and, for those issues considered to be important, the teams define actions and identify responsibilities to provide solutions to these potential problems.

Similarly, for complex or large systems, it may also be worthwhile to undertake a repair and maintenance review. The purpose of this is to identify the potential equipment failures within the system and the processes required to conduct the repair and maintenance necessary to ensure the system continues to operate effectively. This may lead to design changes so that repairs can be conducted quickly, minimising downtime.

Having completed the design and achieved customer approval, the vendor then sources the equipment and commences manufacture of the elements of the cell. In many cases, the system is built on the vendor's premises and fully programmed to enable the FAT to be performed. The customer should visit the vendor's site on a regular basis to view the progress of the work and assess this against the timing plan.

As the project build is approaching completion, the customer must provide parts for the robot programme development, testing, and FAT. These should be truly representative of the actual parts to be processed in production. This can sometimes be difficult to achieve, particularly for new products. In such a case, the customer and vendor must reach a mutual agreement as to how any variances are to be accommodated or assessed.

The FAT is an important milestone. There is often pressure at this stage to ship and install the system because the project is running close to the plan or is actually late. If the system fails the FAT, it is much easier, and less costly, for the vendor to fix problems at its own site. It is also preferable for the customer to delay shipment until the system is proven, because any issues on the customer site can reduce the credibility of the system, which will cause problems in production later. It is therefore better for both parties to ensure the FAT is achieved successfully even if this requires a delayed installation.

8.4 INSTALLATION AND COMMISSIONING

The customer should plan the delivery and installation, with advice provided by the vendor. It may be necessary to decommission and remove existing equipment from the area or to move existing facilities to provide access to the area within the factory where the automation system is to be sited. If an area of production is to be out of operation for a period, it may be necessary to build additional product in advance to ensure production can continue within other areas of the factory. At an early stage in the project, the vendor should have visited the customer site to review the area for the installation and access to the site, and the vendor and customer must devise a joint plan to carry out this work.

If the delivery takes place during production, other operations may be disrupted, and therefore, the production staff need to be made aware of what is required and when this will take place. To avoid this, delivery can take place outside of production hours, either during planned shutdowns or over weekends. It is also important that the customer provides the necessary lifting equipment, such as forklift trucks, unless this forms part of the contract with the vendor. The customer is normally responsible for providing the services, electrical, compressed air and water, if required, to an agreed point within the area of the installation. This work should be completed prior to installation taking place.

There should be a clear definition of work to be conducted by the customer and work to be conducted by the vendor. Related to this division of labour, the customer should ensure the vendor is aware of any supervision or safety issues related to the tasks to be performed. For example, any cutting or welding may require the presence of the site fireman. It may also be necessary for the vendor's staff and subcontractors to undertake a site safety course prior to working on the customer's site, which should be planned in advance.

If possible, the project can often benefit from the customer allocating one or more engineers to assist with the installation, particularly in terms of liaising with local staff to resolve any questions or issues raised by the vendor. These engineers also have the opportunity to be more involved and gain knowledge of the system during this phase, which could be beneficial when the vendor has completed the work and left the site.

Once the installation and commissioning are complete, the SAT is undertaken. This may require an interface to be made between the new automation system and current production, both upstream and downstream. This may cause some issues with production. It is therefore beneficial to review the requirements and tests to be undertaken within the SAT with the relevant production staff.

On successful completion of the SAT, the vendor hands the system over to the customer, and the vendor's staff normally leaves site. The operation of the system is now the responsibility of the customer. In some cases, standby cover is provided to assist with this handover. The need for standby cover depends on the complexity of the system and the existing customer expertise. The purpose of standby is not to operate the system but to provide assistance and guidance in the operation and maintenance of the system for the new operators and maintenance personnel. Therefore, standby is a backup.

During the commissioning period, the vendor normally provides training in the operation and maintenance of the system. This may be prior to the SAT, but it could also be afterwards, once the system is in operation. It is important that the customer and vendor agree on the extent of the training and the numbers of personnel to be trained. This is particularly true if there are multiple shifts to be trained. It is important that the appropriate personnel on each shift have the training necessary to correctly operate and maintain the system.

Documentation is important and should be provided as early as possible to the customer following the SAT. However, the documentation often cannot be finalised until the SAT is completed, in case any further required changes lead to modifications to the documentation. It may be beneficial for the customer to have an incomplete version of the documentation prior to the SAT, partly to enable checking of the information contained, but also to provide some backup once the vendor's staff leaves the site. The final version of the documentation should then be provided promptly once the project is completed.

8.5 OPERATION AND MAINTENANCE

Once the system has been handed over and all of the vendor's staff has left the site, the operation of the system is the sole responsibility of the customer. It is important that the system is operated as intended, and the parts provided to the system must meet the standards specified at the start of the project. In addition, if maintenance is required, such as changing consumable items, then this should be performed at the appropriate frequency using procedures detailed in the documentation.

Even if the system is well-built and fully proven, problems can still arise at this point. These problems are normally due to the inexperience of the customer's operators and engineers who are unable to operate the system as well as the vendor's staff. The problems can lead to the customer requesting support from the vendor because of perceived problems with the system. This situation can often cause conflict, partly because the vendor believes it has completed the work as required and partly because the vendor's staff previously involved in the project may well already be committed to other work.

The provision of the final documentation package can also prove to be a problem because the engineers concerned are often required for other projects. The finalisation of the documentation can therefore become a lower priority, but it is important this is completed quickly. The project is not fully delivered until the documentation package is completed and accepted by the customer.

In most cases once the system is operating as planned, the customer does not consider any changes until a new product or product redesign is to be introduced to the system. However, it may be possible to increase the throughput of the system to gain further capacity. During the project, the vendor sets and achieves a target cycle time. Once the target is achieved, the vendor does not do any further work to improve the output. Over time the customer's staff will have the opportunity to view and assess the operation of the equipment. If they have had the necessary training and are given the opportunity, they may be able to improve the operation of the system. Major changes involving further investment are not normally considered unless a very significant and justifiable improvement can be made. The ad hoc improvements are more likely to be software changes, as with the robot programme, or simple modifications to elements of the system that can be achieved in-house or for very little cost. The gain of production capacity may not be immediately important, but it may well provide benefits in the future and is therefore worthwhile.

The maintenance and service support for the system must also be considered. Normally, the vendor provides a 1-year warranty, but in some cases, the warranty may be 2 or 3 years. Although the vendor will fix problems within the warranty period, this does not necessarily guarantee a quick response to breakdowns. The type and level of support required depend on the expertise of and the training provided to the in-house maintenance staff. If the system is critical to the production of the customer, then a service response package may be necessary, possibly including 24-hour support and guaranteed call-out response times (see Section 6.2.14). This may be appear to be expensive, but compared with the problems it is designed to solve, it could be a worthwhile investment. Even if emergency cover is not required, the customer may want to consider an annual maintenance contract to cover any specific maintenance required, as well as a check of the system. These considerations very much depend on the experience and expertise in-house. If the customer already has a number of robot systems, the operation may be able to handle most of the maintenance and service requirements on its own.

8.6 STAFF AND VENDOR INVOLVEMENT

In most cases it is better to involve staff in projects to gain the benefit of their knowledge and experience, and a number of different groups can make significant contributions to the success of a project. The main obstacle to early involvement is the need to address the insecurity that the introduction of a robot system can raise, particularly in relation to the job prospects of the production staff. If the system is for new work, then it is unlikely that any staff will be displaced, but they may see the introduction of a robot as a threat to their futures. If the system is to take over existing operations, the workers who are performing those operations obviously see the robot system as a threat to their livelihoods. It is important that the rationale for introducing automation is clearly explained to the workforce and the impact on roles and specific jobs is addressed. The management must clearly state whether the affected staff are to be retrained to operate the robots or are to be moved to other jobs within the factory. Whatever the message and however it is delivered, all the parties involved in the automation project must be fully supportive. These parties can make the whole project easier and more likely to achieve success, or they can cause problems, more from lack of interest, cooperation or involvement in the project than from direct obstruction.

8.6.1 Vendors

If a new product is under design and the intention is to utilise automation for the manufacture of that product, the customer can benefit from involving vendors in the design phase. The vendors are able to provide advice on the ease with which the manufacture can be automated, and they may suggest design changes that can significantly aid the automation. If these changes are proposed early in the design stage, it may be possible to implement the changes at little or no cost. This may make the difference between a product that can be manufactured automatically and one that cannot be automated within a cost-effective solution. Even if the design is finalised, or outside the responsibility of the customer, the customer should discuss the design with the vendor. The vendor may be able to suggest small modifications that may be feasible and could have a major impact on the ease of automating.

8.6.2 Production Staff

The involvement of production staff can be very important for an automation project. It is therefore worthwhile to gain their full support and participation as early as possible. If automating an existing operation, the project team should involve the relevant staff at the concept stage. These staff members are performing the operations, and they understand the difficulties within the operations, the variability of the incoming parts, and the true operations that are performed, including any workarounds that have been developed. This type of information is often not detailed on the production instruction sheets, and it may not be known by the production management or the engineering teams. However, without this knowledge of the real situation, an automation project can face difficulties or even fail. Even if the project team is considering an application for new products and operations, the existing production staff may be able to provide valuable input based on its knowledge and experience of current production. The involvement of the production staff can therefore provide very valuable input to the concept design and also the development of the specification.

Continuing the involvement of the production staff throughout the project can also be worthwhile. Production workers do not necessarily have to attend all meetings, but the team should keep them abreast of progress. If, during the design phase, FMEAs are undertaken, the production staff may provide useful input. The FAT at the vendor's premises is also a good opportunity to introduce the operators to the system, enhancing their confidence with the automation. Once the delivery and installation commence, these operators can be part of the installation team providing assistance to the vendor. This increases their familiarity with the equipment and, if this is building on involvement from an early stage, it should also encourage ownership of the system. This ownership can be the key to a successful system. If the operators want the system to work, they try everything they can to ensure that the system does work. Conversely, if they do not feel part of the project and do not have any ownership of the system, they do not actively support resolution of any problems, instead waiting for others to address any issues. The ownership really comes down to a question of attitude: do the operators want the system to work or not? This may be very important because they can be a key element in determining the success of a project.

It is also important to provide training to the production staff at an appropriate stage in the project. The training on the actual system may take place at either the vendor's or customer's site. Investing in more detailed training (e.g. in robot programming) may also enhance the operators' capabilities and status, encouraging their support for the project.

8.6.3 Maintenance Staff

The maintenance staff can provide valuable input during the development of the specification. These staff members have experience with various items of equipment from different vendors already in use within the facility. They can therefore provide details about the most reliable equipment and the vendors who provide good service for breakdowns, maintenance, and spare parts. This knowledge can be distilled into a preferred vendors list to be included in the specification (see Chapter 6).

The maintenance staff can also be involved in reviewing the concept to gain an understanding of how they would access the system for repairs, offering their input as to the feasibility of the proposed solution from a maintenance viewpoint. Likewise, the maintenance staff can provide useful input to the review of the proposals provided by the various suppliers. Also, if repair and maintenance studies are conducted during the design phase of the project, the maintenance staff can provide useful information.

The maintenance staff should receive training on the system, which is likely to take place at the customer's site. It may be beneficial to involve these staff in the installation and commissioning of the system, in support of the vendors staff, to enhance their capabilities. It may also be necessary to provide specific maintenance training on some of the equipment within the system. If this training is required, it should be conducted prior to but close to the date of installation. The staff then have a better understanding when assisting the installation.

8.7 AVOIDING PROBLEMS

All automation projects generally have the following life cycle:

- Project conception
- Project initiation
- System design and manufacture
- Implementation
- Operation

It is possible for problems or failures to occur at any of these stages, and often, when a problem becomes apparent, the root cause occurred in an earlier stage of the project. In order to avoid problems, the teams must invest the appropriate time, resources and expertise into the project at an early stage. For example, detailed investigation in the concept stages provides a better understanding of the risks and potential problems, and therefore, it allows for actions to be taken to avoid these risks and minimise the problems. Notably, the cost required to rectify a problem or to resolve an issue normally increases as the project proceeds. An issue may cost a small amount to fix at the design stage, but the cost to rectify the same issue will be significantly increased if it does not come to light until the commissioning.

This section discusses a number of issues often found within automation projects, and it identifies how these issues might be avoided (Smith, 2001). The intent is to demonstrate how appropriate actions in the early stages of a project can avoid these problems and, therefore, why investment in the early stages of a project is worthwhile.

8.7.1 Project Conception

Project Based on an Unrealistic Business Case

The customer has built a justification for the project based on various assumptions. If these are found to be unrealistic, which may be due to the inexperience of the customer, the team should identify the problem and rebuild the business case based on more valid assumptions. Otherwise, the customer is disappointed at the end of the project, which is not beneficial for the vendor or the customer.

Project Based on State-of-Art or Immature Technology

This problem is relatively common with automation projects. The project initiator is over ambitious in terms of what the company can achieve and what current technology is available and affordable. The initiator should engage the appropriate expertise, even if this requires an external resource. If the only option is to apply unproven solutions, then the budget and timescales must include an allowance for the risk involved. In order to avoid later disputes, the company concerned, particularly the senior management, must be aware of the true risks prior to project commencement.

Lack of Senior Management Commitment

This issue can cause problems later in the project if the managers concerned have influence over one or more of the customer staff involved in the project. For example, if production management are less than supportive, the implementation and operation phases can be very difficult, even if the engineering management fully supports the system. Although difficult to address, vendors benefit from ensuring that the project has the full support of the customer's staff. It is also beneficial for the project engineers, who are responsible for the project, to acquire support from all parties who may have some influence over the project during its complete life cycle.

Customer's Funding and/or Timescale Expectations Are Unrealistic

When competing for a project, a vendor might find it difficult to disagree with a potential customer's expectations for price and delivery. However, it is better to be honest up front and risk losing the business than to accept unrealistic challenges and then fail to deliver later. Provided the vendor's position is justifiable, the straightforward and honest approach may also enhance the customer's confidence in the vendor, resulting in the winning of the order and a better relationship with the customer.

8.7.2 Project Initiation

Vendor Setting Unrealistic Expectations on Cost, Timescale or Capability

The vendor might be very keen to win the business or might lack understanding about what may be involved in executing the project. These situations often result from a lack of experience in the vendor's staff. It is not in the customer's interest to place business with a vendor that has incurred unknown risks or maintains unrealistic expectations in order to win the business. The customer can assess this type of risk by obtaining and comparing multiple quotes, and by reviewing the expertise and resources of potential vendors, including assessing their previous experience via visits to reference sites.

Customer Failure to Define and Document Requirements

The lack of a specification (URS) is the most obvious example of this problem. In this case it is in the interest of the vendor to set out the assumptions behind its offer and to review these assumptions carefully and in detail with the customer. In effect, the vendor-generated assumptions then become the specification, providing the basis against which the project is judged.

Failure to Achieve an Equitable Relationship

The relationship between the customer and the vendor does not need to be friendly, but it does need to be open, equitable and based on mutual understanding. If the relationship is strained in the early stages, larger problems can develop as the project reaches a conclusion. If the personnel cannot be changed or no way can be found to improve the relationship, it may be better for the vendor to withdraw from the project.

Customer Staff's Lack of Involvement

In these cases, the eventual end-users of the automation, such as the shop floor staff, have not been involved in project development and implementation. Although this lack of engagement is difficult address, it is in the vendor's interest to encourage the customer to involve these end-users early on and to encourage their input and support for the project.

Poor Project Planning, Management, and Execution

This can occur on the customer or the vendor side, and it usually manifests as staff in a "fire fighting" mode. The normal cause is the definition and/or acceptance of unrealistic timing plans that do not correctly reflect the scale of the tasks involved or provide any contingency for risks. The problem may not be immediately solvable on a specific project, but lessons should be learnt and investments made, including training if appropriate, to improve for future projects. The costs of running late or over budget are often higher than the training cost incurred.

Failure to Clearly Define Roles and Responsibilities

This issue often arises between the customer and vendor when they share joint responsibilities for delivering the project. It is even more important when the project involves a number of vendors, all of whom are contracted directly by the customer to deliver elements of the project. The difficulties of dealing with multiple vendors leads many customers to insist on one lead vendor that has overall responsibility and project control, with all the other vendors contracted to them, a so-called turn-key project. This arrangement ensures one point of contact for the customer, but it also shifts the responsibility for ensuring appropriate demarcation to the main vendor.

The problems that emerge are gaps in the scope of supply or misunderstandings over interfaces. These often do not become apparent until later in the project, typically during installation and commissioning. Avoiding these issues requires a clear overall specification and very precise definitions of the roles and responsibilities of each vendor.

8.7.3 System Design and Manufacture

Failure to "Freeze" the Requirements and Apply Change Control

The customer may not have provided a specification or clearly defined the requirements, with various iterations also occurring during the quotation process. This dynamic may continue during the initial stages of the implementation. The vendor can easily continue to follow these changes until the cost implications are noticed at a late stage. The vendor then normally tries to recover these costs, which often causes problems between the customer and the vendor. It is much better to fix the specification and then document any changes as they are requested. At the same time, the vendor should notify the customer about the cost implications of these changes, providing the customer the option of reconsidering if the cost is higher than envisaged.

Vendor Starting a New Phase Prior to Completing the Previous One

The vendor can try to recover lost time or accelerate the project by running design and manufacture in parallel. This approach can be dangerous, and it should only be applied if the implications are known and the risks minimal. For example, some manufacture may commence before design is fully completed, if any changes in the design at that stage would have no implications on the manufacture already underway.

Failure to Undertake Effective Project Reviews

If the project proceeds without regular reviews and communication between the vendor and customer, problems are likely to emerge at a later stage. This may be a case when either party takes a "head in the sand" attitude regarding the avoidance of problem resolution. If problems are allowed to continue into later stages of the project, the ultimate resolution will only be more difficult than it would have been if the problem had been addressed earlier.

8.7.4 Implementation

Customer Failure to Manage the Changes Implicit in the Project

Particularly for a first project, the introduction of automation may require changes to the attitudes and possibly the working practises of some customer staff members. The system will only work successfully if those who operate and maintain the system wish it to work. Therefore, the customer must gain the support of all those who become involved. This should be done prior to implementation, and, if the vendor has concerns in this respect, it is worthwhile to raise these with the customer at an early stage in the project.

8.7.5 Operation

Inadequate User Training

This often becomes apparent when the vendor leaves the site, and the system does not then operate with the same performance or reliability as it did when the vendor was present. The vendor must ensure that the customer's staff receives the appropriate training so it can operate and maintain the system as required.

Customer Fails to Maintain the System

Although difficult to influence, the vendor must ensure that the customer understands the need for appropriate maintenance if the system is to continue to operate as intended.

Customer Fails to Measure the Benefit of the Project

The project engineers who initially developed the business case for the project should ensure that the actual performance and output are measured to compare with the justification used to gain approval for the project. Any over performance should be highlighted, and any under performance investigated to provide guidance for future project justifications.

8.8 SUMMARY

When undertaking automation projects, the most common mistakes made by customers include a lack of control of the input parts, failure to gain buy-in from the shop floor, failure to explain what is really required to the vendor and finally the selection of the vendor based purely on the lowest purchase cost. All of these mistakes are avoidable, however, given the appropriate approach to the project.

Automation highlights any input quality problems because it does not have the same flexibility as a manual operation. However, provided the parts input to the system can be controlled to achieve the parameters against which the system was designed, the overall quality of the output will improve.

The development of a detailed specification provides the basis for communicating the system requirements to the vendor. Without this specification, the vendor has no formal explanation of what is required, which can, and often does, lead to misunderstandings and disagreements later in the project. Therefore, this is an important document, and the customer should invest the time and resources required to provide a comprehensive specification (URS).

It is not surprising that the selection of the vendor is also critical. Yet, the customer often selects the lowest-cost vendor, with very little investigation of their capabilities. In such cases, a specification can provide some protection, but if the vendor fails to perform, the customer still incurs added cost. Again, the customer should invest the time required to fully investigate vendors and their expertise in order to ensure that the vendor selected is the most appropriate one for the project, even if that vendor did not offer the lowest quote.

A key part of any project is the two-way communication between the customer and vendor. The customer must explain the basic details of its requirements, the specification, and also cover the actual input tolerances, the details of the true operations, what is likely to go wrong and the flexibility needed. In turn, the vendor should explain the true production rate, including availability, the ramp-up time, the skills required to run the system, and the acceptable input tolerances. This needs to part of an open relationship that is not solely based on cost but also values expertise and experience. With a partnership between the vendor and customer, it is possible to overcome many problems, and the final results will be beneficial to both parties.

The customer must also encourage their staff to become involved, and it must provide training for the different people who will work with the system. This involvement generates familiarity with the equipment, removing the fear factor. It also gives the team the ability to quickly correct problems, and it creates ownership of the equipment, providing a desire to fix problems and ensure the system runs as planned. It also gives the customer the confidence to undertake improvements after the vendor has left, which may lead to greater output than planned. These benefits outweigh the cost of the training, which is therefore very worthwhile.

Finally, the actual performance and savings resulting from any automation projects should be compared to the original justification, and the final costs should be compared to the budget. This review indicates if the project met its objectives, and it also provides valuable lessons for future projects.

CHAPTER 9

Conclusion

Chapter Contents

Abstract

This chapter brings together the automation equipment, applications, and project stages discussed in the previous chapters. The development of an automation strategy is discussed, including the relationship between automation and lean manufacturing. The benefits of such an automation strategy are identified and discussed. The chapter concludes with a view of the future together with some key points all manufacturing companies should consider in relation to their investment plans for robot systems.

Keywords: Lean manufacturing, Automation strategy

Early automation systems were dedicated to specific products. These were suited to high volumes with a limited range of product variants and were expensive to modify for changes in products and volumes. The advent of robots has provided the opportunity to develop and implement flexible automation systems that are better suited to current production requirements; that is, the ability to cater for a larger range of product variants, shorter product life, and also variation in volumes.

Robots are a form of automation that embodies the highest levels of flexibility. They were initially seen as having the capability to be reconfigured not only for different products but also for completely different applications. Robots still have this capability, although they are not often reconfigured in this way. With the exception of some specialist applications such as painting, a particular design of robot can be applied to a range of applications. The basic configurations of robots and the key issues involved in selecting the most appropriate machine for a particular requirement were reviewed in Chapter 2.

Robot software has become increasingly important. The enhanced capability of current controls has provided significantly improved robot performance and has made the systems much simpler to develop, install, and operate. Dedicated application software has become an increasingly important element of the system, providing the ability to configure and control application equipment as well as operator interfaces specifically designed for the application.

As robot technology has been enhanced, additional devices have been developed as standard products, such as positioners and tracks, to provide the robots with increased capability for specific applications. In addition, function packages have been developed to provide application capabilities, fully integrated with the robot, both mechanically and as part of the control functionality. These function packages reduce the need for the design and development of bespoke solutions and thereby reduce the time required to produce a robot system as well as providing more cost-effective solutions and increasing the reliability of the robot system.

Robots have become proven elements of a flexible automation solution. However, it is important to note that the robot is only one element of a complete solution and on its own can accomplish very little. A complete system requires other equipment to provide the robot with the capability to integrate within the overall manufacturing operation and perform the required tasks. Some of the more common elements of an automation system were reviewed in Chapter 3. The applications of robots to the more common applications and the specific needs of these applications were reviewed in Chapter 4.

The development of an appropriate concept for a robot solution to a manufacturing problem may initially appear complex but it is really a set of steps that are no different than any other capital investment project; therefore, it can be handled by any competent engineer. First, the requirements of the project must be defined including parts, tasks, and production rate. It is then possible to define the main elements of the system, including the selection of the robot type and the numbers of robots required. The basic layout can then be configured, including the part in-feed and out-feed as well as the operator interface and the outline of the safety system. This process was discussed in more detail in Chapter 5.

At this stage the financial aspects of the project can be considered both in terms of the costs for the implementation of the concept and also the cost benefits that may accrue from its implementation. The cost-benefit analysis will then determine if the project may be viable or should be reconsidered to improve the likely justification. These initial stages for the justification were reviewed in Chapter 7.

If the project appears to be viable, more detailed work can be undertaken. This may well involve external suppliers and, at this point, it is important that the detailed specification be prepared to ensure there is a clear and common understanding of the requirements for the project. This was discussed in Chapter 6, and it should again be stressed that the time invested in the specification is very worthwhile and can avoid significant problems and potential costs later in the execution of the project.

The final internal hurdle before a project is given approval to proceed is almost always the financial justification for the project. All companies require some form of return for any investment, and all projects must meet this requirement. It is again worthwhile investing time during the development of the financial justification to ensure all potential benefits are identified and expressed in financial terms (see Chapter 7). This work will ensure the appropriate budget is available, which will in turn enable the purchase of the necessary equipment and services to achieve what is planned.

Although appropriate financing is a key element to any project, it is also necessary to execute the project correctly to ensure all the work on the equipment selection, concept development, and financial justification does achieve the required result. The key elements of this process, including some problems that can be avoided, were discussed in Chapter 8. If the appropriate budget has been defined, this will enable the purchase of the correct equipment and necessary services. If the project is then executed correctly with the appropriate planning and interaction between supplier and customer staff, the resulting system will then perform as required. As a result, the benefits will be produced as anticipated and the planned financial returns will be achieved. The project will then be a success.

It is risky to shortcut any of these stages. Without detailed work in the early stages of any project the risk of unforeseen issues increases and the impact both on the cost and timing of a project increases the longer these issues remain. So, it is worthwhile investing time, resources, and sometimes money to ensure all risks are understood early and solutions to those risks are identified. Likewise, detailed study to ensure the financial analysis is correct will also be highly beneficial. In some cases projects may not proceed, due to technical risk or the lack of an appropriate financial return. These projects may not be viable in the short-term, but over a couple of years the cost-benefit may improve or new solutions may become available. It is better to wait and achieve a successful project in the future than to execute a project that is destined to fail in some way.

9.1 AUTOMATION STRATEGY

All the above steps are a necessary part of the conception and execution of every robot project. It is possible to bring in external resources to assist at various stages. This may be via suppliers or, alternatively, independent consultants who may provide a more balanced approach as they have no specific interest in any decisions made at any stage. However, it is better to enhance the capabilities of the internal staff so they are able to perform this work independent of external support. Training of staff, both for the initial stages of the project and also the operation and maintenance of robot systems, is very beneficial. It provides greater self-sufficiency and, therefore, reduces the likelihood and impact of any problems. This investment in staff is not often achieved if projects are considered only as one-offs. A more strategic approach to automation is required.

Most organisations have developed and implemented business plans. These consider issues such as product development, growth targets, and financial resources. To achieve the best from automation it is beneficial to develop and implement an automation strategy. This may form an element of the overall business plan or could be a stand-alone strategy for the improvement of the manufacturing operation.

One key aspect of any manufacturing operation is the effectiveness and efficiency of the activities within the operation. Lean manufacturing as a philosophy was initially developed in Toyota. The objective of lean manufacturing is to improve the operation by focusing on the seven manufacturing-related wastes (Womack and Jones, 2003):

1. Overproduction.
2. Excessive movement throughout a process that is not required to build an item.
3. Delays between production steps.
4. Excess inventory.
5. Excessive movement of people or equipment that is required in the processing of a part.
6. Overprocessing of parts.
7. Finding and fixing defects.

Automation may assist in addressing some of the above wastes but conversely, if badly planned or implemented, it can actually increase the waste and thereby negatively impact profitability. Therefore, it is always worthwhile to consider the basic principles of lean manufacturing when developing an automation strategy. In many cases it may be beneficial to embark on the lean journey and make initial progress before considering any form of automation. There may well be limited value in the automation of one specific step in the manufacturing process, particularly if all this does is generate greater problems elsewhere.

The real benefits of automation will be realised if the overall manufacturing strategy encompasses both lean manufacturing as well as a longer-term automation strategy. This automation strategy should include a longer-term plan, possibly as many as 10 years hence, with a goal defined as to what the manufacturing operation might look like at the end of that period. This may be considered too difficult or subject to too many variables, including technology changes, to be worthwhile. But if you do not know where you are going, how will you get there? If it is possible to develop a business plan based on an assessment of the future, it is also possible to achieve the same for manufacturing and the automation to be used within it.

An automation strategy provides the opportunity to address longer-term issues that often impede the implementation of automation. The first of these is training and skills of both the workforce and the engineering teams. A strategy assists in defining the longer-term needs and providing a direction for the training to achieve these needs. First, the engineering resource tasked with identifying opportunities for automation and then developing concepts and implementing the automation solutions will benefit from training on project implementation and management, as well as more specialised training on automation and robotics. This resource would also benefit significantly by removal from production activities and given the time to develop their expertise and proposals without the pressure of normal day-to-day activities. This investment in time and focus on a small group of staff, perhaps a single person, will pay dividends in the long run.

The shop floor also benefits from some training. In particular the supervision and maintenance staff can benefit, as it is their role to ensure that the automation continues to operate once installed. There is often merit in training a small number of staff to become the machine minders to take responsibility for the day-to-day operation as well as programming for future products and also enhancement of the existing systems. With the correct skills these staff can become key employees ensuring the best output from existing automation as well as contributing to the development of new systems.

It is beneficial to communicate major changes to all production staff. Automation can be seen as a threat to jobs as it often does replace some job functions. This can be achieved on a project-by-project basis but is much easier to achieve if there is a longer-term strategy that demonstrates how the business intends to grow. Therefore, the automation, rather than being a threat, is seen as positive both for the current staff and future employment.

The second main hurdle, which an automation strategy can help to address, is that of financial justification. Many projects do not get approved because they do not meet the financial criteria required. In some cases, simple projects cannot be implemented for the same reason. Many companies have had to undertake more complex projects because these were the only ones that could meet the payback requirements of the business. This approach causes two problems. First, projects that would provide benefit are not being implemented. The payback may be longer than required but if the equipment will provide value for a number of years, often over 10 years, than the business would be improved as a result of the implementation. Second, it is much easier and less risky to implement the simple projects, which are often much more likely to succeed. They also provide valuable learning opportunities for the staff, whom can then move on, having gained the necessary expertise, to more complex projects. For these reasons, an automation strategy provides a vehicle by which the business can be improved, having convinced senior management of the longer-term benefits of automation and the need to take appropriate steps towards achieving the end goal.

Related to the finance issue, if the business is seeking external financing to implement an automation solution, it can be beneficial to be able to demonstrate that the investment forms an element of a longer-term plan. Financial institutions are much more likely to provide financing to companies that can demonstrate they have a coherent plan. An automation strategy, demonstrating how the business intends to improve competitiveness and profitability, can be a valuable aspect of such a plan.

An automation strategy can also be beneficial when competing for new business with either new or existing customers. If the company is able to demonstrate how it is planning to improve its manufacturing operation over the coming years it is more likely to win business, possibly at a price premium, particularly if the company can also demonstrate how its customer will benefit. Many companies are faced with the need to implement automation systems when they have just won a new job. This can be achieved if the company has the necessary experience and can execute a project quickly. It can be very challenging for those companies undertaking automation for the first time. It is, therefore, better to already have experience with automation and be prepared, rather than to attempt to install automation once a job is already won. An automation strategy can address this issue, particularly if it is developed in conjunction with the overall business plan.

A further aspect of such a strategy is the opportunity to develop long-term partnerships with suppliers. Understanding the longer-term needs of the company, rather than simply selecting suppliers on a project-by-project basis, provides the opportunity for much stronger relationships and mutual understanding. This, then, delivers better results for both parties. Communicating some aspects of the strategy to key suppliers provides them with an understanding of the company's future needs and potentially helps them ensure they are prepared to meet those needs, in terms of products and services, as and when they are required.

An automation strategy can be very beneficial. It must be closely aligned with the overall business strategy and so does need the agreement of senior management. This may take time to develop but once completed the clarity of the longer-term plan and the steps to be implemented make it much easier for a company to develop their automation on a step-by-step basis, improving skills, expertise, and competitiveness as the plan develops. Similar to any business plan, the automation strategy should be reviewed on a regular basis to ensure it meets the needs of the company and also to ensure it takes account of technology changes.

9.2 THE WAY FORWARD

At the time this book was written, in 2014, the industrial robot was, in the main, a single-arm device and needed to be isolated from manual workers with safety guarding. This situation was about to change with the imminent launch of dual-arm robots that could work alongside and in collaboration with humans. It was anticipated that this could herald a new era for industrial robots, opening up new applications, particularly in assembly applications and operations. There has also been significant interest in the development of robots that are easier to use and programme, particularly for smaller businesses. Both these developments could have a significant impact on the scale of the industrial robot market, particularly if the overall system pricing is not dissimilar to the cost of hiring workers.

There continues to be significant developments in the service robot sector. Robot technologies continue to be developed to provide greater capability to handle unstructured environments, to provide safe interaction with humans and to improve ease of interaction with humans. The developments within the service robot sector will be of benefit to the industrial robot sector; partly from the development of new technologies, but more importantly from the scale that will be achieved from the commercialisation of these technologies. Service robots, for the general consumer, will be sold in much larger volumes than industrial robots and, therefore, the production of these machines and devices will provide significant economies of scale and lower costs. If the same technologies can be applied to industrial robots, there will be significant cost-benefit for users of industrial robots.

These developments could lead to a step change in the number of robots being sold into the manufacturing sector. The benefits of flexible automation will include increased competitiveness and profitability for any company that has taken the appropriate steps to utilise robots within its manufacturing operations. Such companies will succeed and grow. It is, therefore, important for all manufacturing businesses to investigate robot technologies, to assess the suitability for their manufacturing operations, and to develop a strategy to implement flexible automation solutions where appropriate.

Taking a lesson from Henry Ford:

> *If you need a machine and don't buy it, then you will ultimately find that you have paid for it, but don't have it.*

Basically, if you are spending the money anyway you should buy the machine. With flexible automation this sometimes requires longer-term thinking but in the long run provides the best result for the business.

Also a word of warning from John Ruskin:

> *It's unwise to pay too much, but it's worse to pay too little. When you pay too much, you lose a little money – that's all. When you pay too little, you sometimes lose everything, because the thing you bought was incapable of doing the thing it was bought to do.*
>
> *The common law of business balance prohibits paying a little and getting a lot – it can't be done. If you deal with the lowest bidder, it is well to add something for the risk you run, and if you do that you will have enough to pay for something better*

Please remember the cheapest solution is not necessarily the best, and always select suppliers very carefully with detailed reviews of their proposal, experience, and expertise.

Finally, a plea to all manufacturing companies that have yet to fully embrace the benefits of robots and flexible automation. I believe there are three pillars which form the foundation of a successful manufacturing business:

- Product and process innovation.
- Effective organisation (lean manufacturing).
- Capital investment in equipment.

The best and most competitive businesses have considered all three pillars and have plans in place to continually improve all three aspects of their business. Each one is important, but the pillar that often has the least attention is the investment in equipment. In today's world, flexibility is key and, therefore, robots should form an important element of every manufacturing business.

REFERENCES

Engelberger, J.F., 1980. Robotics in Practice. Kogan Page Limited, London, UK.

International Federation of Robotics and United Nations: Economic Commission for Europe, 2005. World Robotics: Statistics, Market Analysis, Case Studies and Profitability. United Nations: International Federation of Robotics, New York.

International Federation of Robotics, 2011. Positive Impact of Industrial Robots on Employment. Metra Martech Limited, London, UK.

International Federation of Robotics, 2013. World Robotics, Industrial Robots 2013. International Federation of Robotics, Statistical Department, Frankfurt, Germany.

Smith, John M., 2001. Troubled IT Projects Prevention and Turnaround. Institution of Electrical Engineers, London, UK.

Womack, J.P., Jones, D.T., 2003. Lean Thinking: Banish Waste and Create Wealth in Your Corporation. Simon & Schuster, New York.

ABBREVIATIONS

2K	two component
3Ds	dirty, dangerous, and demanding tasks
AGVs	automated guided vehicles
CAD	computer-aided design
FAT	factory acceptance tests
FDS	functional design specification
FMEA	failure mode effect analysis
GPS	global positioning system
HMI	human machine interface
IFR	International Federation of Robotics
IP	ingress protection or international protection
IP67	Ingress protection; 6 = totally protected against dust, 7 = protected against the effect of immersion in water between 15 cm and 1 m deep
I/O	inputs and outputs
MAP	manufacturing automation protocol
MES	manufacturing execution systems
MIG	metal inert gas
MIT	Massachusetts Institute of Technology
MTBF	mean time between failures
MTTR	mean time to repair
PLC	Programmable Logic Controller
PPE	personal protective equipment
PUMA	programmable universal machine for assembly
RSI	repetitive strain injuries
SAT	site acceptance test
SCADA	supervisory control and data acquisition systems
SCARA	selective compliance assembly robot arm
SOP	start of production
TIG	tungsten inert gas
URS	user requirements specification

BIBLIOGRAPHY

The following Web sites provide useful sources of information. These include industry associations as well as some of the relevant robot and equipment suppliers.

Associations

International Federation of Robotics. www.ifr.org.
Robotic Industries Association (USA). www.robotics.org.
British Robot and Automation Association. www.bara.org.uk.

Robot Manufacturers

ABB. www.abb.com/robotics.
Fanuc. www.fanucrobotics.com.
Kuka. www.kuka-robotics.com.
Yaskawa. www.yaskawa.eu.com.
Kawasaki. www.khi.co.jp/english/robot.
Toshiba. www.toshiba-machine.co.jp/en/product/robot/.
Mitsubishi. www.mitsubishielectric.com/fa/products/rbt/robot/.
Adept. www.adept.com.
Staubli. www.staubli.com/en/robotics/.

Automation Equipment Suppliers

SVIA (machine tending). www.svia.se.
Schunk (grippers). www.gb.schunk.com.
Unigripper (vacuum gripper). www.unigripper.com.
RNA Automation (feeding equipment). www.rnaautomation.com.
Cognex (vision). www.cognex.com.
Festo (automation devices). www.festo.com.

APPENDIX

Examples of:

Request for Quotation Letter

User Requirements Specification

(To be on Company headed paper)

Dated *xx/xx/xx*

Dear *xxxx*,

Project Name

Please find enclosed our User Requirements Specification for a robot welding system to be implemented at our facility in *(address)*.

We now wish to receive formal quotations for the above project and invite you to submit your detailed proposal. Your quotation should provide sufficient detail to enable a fair assessment of your offer. At a minimum this should include the following:

- Timing plan (from order placement to full production), including significant milestones.
- Outlined system design including a layout drawing and main dimensions.
- Details of the main items of equipment.
- Assembly cycle times, production throughput, and availability.
- Any limitations of the system.
- Details of any proposed subcontractors and the equipment/services they are to provide.
- Any specific floor and foundation requirements.
- Services requirements (including location on the above layout).
- Price.
- Acceptance of the *(company)* Terms and Conditions (attached).
- Any exclusions.

It would be beneficial if you could also include details of appropriate customer reference sites and indicate whether we may contact them.

Please respond by return to confirm your intention to submit a proposal.

The deadline for receipt of proposals is 5pm on *(Date)*.

Following an initial review of the proposals, we will invite selected companies to *(company)* to provide a presentation and enter into a more detailed discussion regarding their proposals. This is planned to take place before *(date)*. We may then wish to visit the premises of those companies and appropriate, existing clients. We intend to make the final supplier selection by the *(date)* and to place the order shortly thereafter. We anticipate that the system will be operational by *(date)*.

The following payment terms are to be applied with invoices to be submitted as follows:

20% on order acceptance
20% after design approval
40% after the delivery of equipment is completed
10% after successful completion of Site Acceptance Tests
10% 30 days after full project completion (including delivery of standby, training, and documentation).

Payment will be subject to the *(company)* Standard Terms and Conditions, which are attached.

If you have any questions regarding the project please do not hesitate to contact me.

Yours sincerely,

(Project Leader or Purchase)
(contact details)

USER REQUIREMENTS
Specification

Project Name	
Revision Number	
Document Status	
Issue Date	

Contents

A.1 OVERVIEW

The customer is a leading subcontract engineering customer with X manufacturing sites in the UK. The customer manufactures a wide range of fabrications, in steel and aluminium, for OEMs in the Truck, Wind Energy, and Construction sectors.

The customer has made significant investments in CNC machinery for the initial preparation of parts, including laser cutting and bending, and therefore manufactures parts to a high standard and tight tolerances. At this time, however, all welding operations, the majority of which utilise the MIG welding process, are carried out manually. The intention is to invest in appropriate automated welding equipment to improve the productivity of the welding operations and the quality of the parts being produced.

This document is intended for quotation purposes. It provides the basic information necessary for vendors to propose solutions and develop proposals. It is expected that a Functional Specification will be developed in conjunction with the chosen vendor once the final solution has been defined.

A.1.1 Current Welding Operation

The customer has a diverse range of MIG welding equipment, including systems from both Esab and Kemppi (generally 300–350 amp). The fixtures have been manufactured in-house and provide simple location of the parts with manually operated clamps included as necessary.

Further details of the assemblies to be addressed are provided in Section A.2. The input parts are prepared in a separate facility, which includes the cutting and bending operations, close to the welding area. The parts are prepared in batches that can vary from 20 to 100 units. These are loaded into various bins and carriers, dependent on the type of part, and transported to the welding area.

Each welding station is separated by welding screens and generally comprises a table on which is mounted the relevant welding fixtures for the part to be processed. Welding fixtures are stored adjacent to the welding area. Each welding station is equipped with one set of MIG welding equipment and is normally occupied by one welder. In total there are 20 welding stations.

The welding facility operates two shifts, 5 days per week, with 8 hours per shift. There are 20 welders employed on days and 10 welders employed on the night shift. The parts for the assemblies to be produced are delivered to the welding station. Each welder will take parts as appropriate, locate them in the fixture, and perform the necessary welding operations. The assemblies are typically produced in stages, although for some of the smaller assemblies all the parts are located in the fixture prior to welding. Once an assembly is completed it is placed in a rack or bin located outside the welding station for removal and replacement when full.

A.1.2 Automation Concept

The customer intends to introduce a number of robot welding systems over time to automate the majority of the welding operations. This will be accomplished in a number of stages that must each be effective and financially viable. The customer is therefore looking for guidance as to how this would be best achieved and the types of automation solution that should be introduced.

This specification provides details of all the parts currently being produced (see Section A.2). It is understood that not all these will be processed by the first robot system. The vendor is therefore requested to consider which parts should be addressed by the first system to be introduced to ensure this first system achieves the objectives for the project, is fully utilised, and provides a sound introduction for robot welding.

The customer envisages an automation system concept that allows the operator to load individual parts into a fixture(s) at one work station, whilst the robot performs welding operations at a separate station. On completion of both operations, the robot and operator will then switch stations to allow the robot to continue welding whilst the operator unloads the welded part and loads new parts to be welded. It will be necessary to balance the welding time and the load/unload time to ensure the utilisation of the robot is maximised as well as ensuring the operator is utilised effectively.

The proposed solution must include the ability to change fixtures quickly and with repeatability to ensure changeover between different assemblies is accomplished with the minimum of down time. The concept must also provide easy access for the delivery of parts and the removal of completed assemblies.

The preference is to apply a standard concept across the full range of current products. The result would be a number of common cells that then have the flexibility to produce any of the assemblies. However, the customer is willing to consider alternative concepts for some parts, if there are significant benefits from this approach. The utilisation of the equipment will need to be more carefully considered if more than one concept is utilised. It is important to note that the automation concept must be flexible to cater for future product design changes and also new products.

The layout of the current welding facility is provided below. The location of the first robot cell is not yet defined, and the vendors are requested to consider how best to accommodate their proposed concept within the existing facility. This should take into account the need to continue with current production during the installation and commissioning phase for the robot cell. The customer recommends that the vendor make arrangements to visit the site to view the area and access for the equipment during installation.

A.2 REQUIREMENTS

This section identifies the main parameters to be considered and the operating requirements for the system.

The solution proposed must be fully automatic once the cycle is initiated, with the capacity to run unattended save for normal operator loading and unloading sequences. The system must be designed to operate reliably on a 24/7 basis with the minimum of intervention for maintenance or fault rectification.

A.2.1 Products

The assemblies currently produced are listed below, and drawings detailing the parts and welding requirements are enclosed.

Name	Part number	Volume PA	Typical batch size

A.2.2 Tolerances and Quality

The tolerances of the parts and also all assembly tolerances are included on the drawings. If the vendor identifies any concerns in relation to the ability to robot weld these parts and/or achieve the assembly tolerances, these must be identified in the proposal.

The weld requirements are identified on the drawings. In addition to meeting these, all welds must provide a good cosmetic appearance with minimal spatter. The customer will provide, to the selected vendor, a set of manually welded parts that will be used to determine the minimum acceptable standard from the robot cell.

The system will be required to achieve the above quality standard with a yield of 99.5%. The yield is defined as the number of acceptable parts produced per shift divided by the total number of parts welded over the same period.

A.2.3 Fixtures

For the first system, and the assemblies selected to be produced by that system, the customer expects the vendors to propose appropriate fixtures and to include these within the scope of supply. The vendors should include the most appropriate clamping techniques and fixture coding to provide for automatic selection of the appropriate robot programme based on the fixture(s) located within the cell.

Changeover of one fixture to another must be fast and repeatable. Vendors should include within their proposal the optimum method of achieving this to suit their chosen concept.

The vendor is to propose those assemblies that provide for the most effective operation of a single robot cell, as well as the most cost effective solution. The customer intends the cell to operate in line with the current operational times of the welding facility, that is, over two shifts.

A.2.4 Cycle Time and Availability

The vendor must provide cycle time estimates at 100% efficiency for the parts they have selected to be produced by the robot cell.

In addition, production outputs should be determined based on the batch sizes and overall volumes identified in Section A.2.1. The target availability for the system is 85%. The production output calculations should include 15% down time due to maintenance, fixture changeovers, and so on.

A.2.5 Welding Equipment

The customer will accept recommendations from vendors in relation to the most appropriate welding equipment for the robot cell. Vendors must ensure that all welding equipment proposed is capable of meeting the power output and duty cycles envisaged for a system of this type and the assemblies and production rates proposed.

The welding equipment should include the capability to weld the material for the assemblies selected. If an equipment changeover is required between assemblies, from steel to aluminium or vice versa, the proposed method of changeover and the time required to perform this must be identified.

A complete welding package must be offered, including torch mounting, torch cleaning, antispatter spray, and wire cutting. Additionally, recommendations on wire delivery, reel or bulk pack, would be welcomed.

A.2.6 Controls and HMI

An operator panel will be provided, mounted at a convenient position adjacent to load/unload area. This is to provide all the functionality to operate and maintain the system, as well as to recover from faults. The minimum functions to be included are:

- Assembly/program selection
- System start
- System stop
- Emergency stop
- Fault location indication

The control system will also provide production management information including:

- Number and type of assemblies produced per shift
- Cycle time
- OEE

The system is also to include a down time log, with time and date, including the cause of the stoppage and the time down until the fault is cleared. This is to cover the main elements of the system, for example:

- Weld equipment fault
- Robot fault
- Stoppage due to lack of product
- Stoppage for cleaning/preventative maintenance.

The robot and/or control system is to include the ability to backup and reload programmes. They are also to have suitable, tiered password protection to prevent unauthorised changes to programmes or settings and records.

A.2.7 Enclosure

The enclosure will be constructed from steel panels with appropriate viewing panels. It must ensure no access is available to the automated equipment during operation. Simple and safe access, using appropriate interlocks, must be provided to all areas of the system to clear problems in the event of a fault. The enclosure will provide protection from the welding arc to both the operator and other personnel in the vicinity of the cell.

The cell must operate at a noise level below 80 dB at 1 m when in automated continuous operation. If the process is generating noise above this limit, the enclosure must include suitable sound suppression to reach the desired limit.

The enclosure will include a fume hood over the welding area, with appropriate lighting. The fume hood will include a suitable port to allow the customer to connect to ducting to remove weld fumes from the cell.

A.3 SCOPE OF SUPPLY

The scope of supply is for a complete, turn-key, robot welding cell and includes the design, manufacture, assembly, testing, delivery, installation, and commissioning of the equipment required to provide quality welds on specified assemblies.

A.3.1 Free Issue Equipment

None.

A.3.2 Safety

The design and manufacture of the automation system must be in accordance with the current safety legislation (see Section A.4.6). When operated and maintained as specified in the documentation (see Section A.3.13) and the training (see Section A.3.14) provided by the vendor, the system must be safe. It must also be safe for any other personnel and untrained operators who may occasionally either use or be within the vicinity of the system.

A.3.3 Services

The customer will provide the following services to one location, to be specified by the vendor, within the vicinity of the system:

- 415 V, three phase
- 24 V dc
- Compressed air, 70 psi

Please note it is the responsibility of the vendor to ensure that the air and power supply have the cleanliness and stability required for consistent and safe operation of the automation system.

A.3.4 Project Management

The vendor will appoint a Project Manager who will take responsibility for the project and liaise with the nominated representative of the customer. The Project Manager's name and contact details will be provided to the customer on acceptance of the order.

A kick-off meeting will take place, within 14 days of order acknowledgement, at the customer and will include the Project Manager, the project team (including representatives of any major subcontractors), and the customer team. This will establish the project aims, schedules, and communication channels and confirm project timescales and the critical path.

Within 7 days of the kick-off meeting, the Project Manager will issue an updated and detailed project timing plan.

The Project Manager will be responsible for the vendor fulfilling its obligations within the contract timescale, including:

- Project planning
- Project resourcing
- Quality control
- Project monitoring
- Subcontractor control
- Customer communications
- Change control

A.3.5 Design

The mechanical design will be developed using appropriate CAD tools and reviewed with the customer prior to manufacture.

A Functional Design Specification (FDS) will be produced for the system and the functionality reviewed with the customer prior to manufacture.

The reviews with the customer are to include a specific review of safety with the customer's Health & Safety staff.

Approval of the design by the customer will not include any acceptance of responsibility for the performance of the equipment, which will remain the sole responsibility of the vendor.

A.3.6 Manufacture and Assembly

Following design and design approval, the manufacturing and assembly will commence. Bought-in items will be procured, mechanical and control system manufacture will commence, and software will be produced.

All assembly work will be carried out at the vendor's premises or, if at an alternative site, the location must be disclosed to the customer. The customer reserves the right to view the manufacture and assembly at any stage subject to the provision of reasonable notice to the Project Manager.

All work and manufactured equipment including software must comply with current regulations (see Section A.4.6).

A.3.7 Pre-delivery Tests

Upon completion of manufacturing, the system will be brought to a stage where its functional operation can be satisfactorily demonstrated. The customer will deliver a quantity of parts for the relevant assemblies to the site, based on timing and requirements agreed upon with the Project Manager, to enable the vendor to test the operation of the system.

The vendor will then carry out Factory Acceptance Tests (FAT) to demonstrate the system performs to the required specification. The customer will be invited to witness these tests. Details of the tests will be provided for the customer's approval, in the form of a Functional Test Specification, prior to the tests being performed.

The FAT will include a preliminary inspection of the equipment and documentation in line with the User Requirement Specification. This will be followed by functional testing as defined in the Functional Test Specification, which will include operation of the cell for a continuous period.

During this period, 5 units of each assembly will be produced, and on completion of these 5 assemblies, the fixtures will be changed. The cell will then continue with the 5 units of the second assembly, until all the assemblies to be produced by this cell have been tested. The Vendor will be responsible for the loading and unloading of the system for these tests.

During this test, the following information will be recorded for each variant of assembly produced:

- Cycle time for each assembly produced (CT1, CT2, CT3, CT4, CT5)
- The length and cause of any down time (DT)
- The time required to perform the fixture changeover (FCT)

Cycle Time and Availability Calculations

The total time recorded for the production of each batch of five assemblies will be compared with the cycle time estimates provided by the vendors for those assemblies at 100% efficiency (see Section A.2.4) to determine if the cycle time estimates have been achieved.

That is

$$\text{Cycle time} = (\text{CT1} + \text{CT2} + \text{CT3} + \text{CT4} + \text{CT5})/5$$

The actual time to produce the specified batch sizes (see Section A.2.1) will then be calculated based on the total time for the 5 units of each assembly to give a figure "welding time".

That is

$$\text{Welding time} = \text{Cycle time} * \text{Batch size}$$

The causes of the down time will be reviewed to determine if they are due to the function of the equipment or other issues outside the performance of the cell. Any revisions will be incorporated with the down time (DT) figure. This revised figure will then be multiplied by the ratio of the actual batch size (Section A.2.1) to the 5 units to give a figure "down time".

That is

$$\text{Down time} = \text{DT} * \text{Batch size}/5$$

The production time for one variant of assembly will then be:

$$\text{Production time} = \text{Welding time} + \text{FCT} + \text{Down time}$$

The above calculations will be repeated for each of the assembly variants (assy 1, assy 2, . . .). The total welding time will be the sum of the welding times for all the assembly variants.

$$\text{Total welding time} = \text{Welding time (assy 1)} + \text{Welding time } (assy\ 2) + \cdots$$

The total production time will be the sum of the production times for all of the assembly variants.

$$\text{Total production time} = \text{Production time (assy 1)} + \text{Production time (assy 2)} + \cdots$$

The availability will then be determined:

$$\text{Availability} = \text{Total welding time}/\text{Total production time}$$

The availability will then be compared with the target specified in Section A.2.4.

If the FAT does not meet the vendor's specification, the vendor reserves the right to request that modifications or rectification work is performed and the FAT is repeated prior to delivery of the system.

A.3.8 Delivery

Once acceptance tests have been satisfactorily completed, the equipment will be ready for despatch and delivery arrangements may be made.

The vendor will be responsible for returning to the customer, at their cost, the assemblies produced as a result of these acceptance tests, together with any unused parts.

A.3.9 Installation Requirements

At an early stage in the project, the vendor will conduct a site survey. The purpose of the survey is to check the floor, confirm building dimensions and the positions of adjacent equipment. It is not envisaged that any foundation work or floor modification will be required, but it is the vendor's responsibility to notify the customer of any requirements early in the project.

The vendor will produce layout drawings confirming the proposed position of the equipment. The vendor will also indicate the required position for the services (air and electricity). These must be approved by the customer prior to delivery.

A.3.10 Installation and Commissioning

The Project Manager will provide a clear written "statement of works" for the installation of the cell, including a safe system of work, having undertaken a risk assessment beforehand. The installation plan will also identify the number of personnel involved, any assistance required from the customer and also any specific requirements that may interrupt existing production.

The vendor's engineers will install the equipment at the site. The customer will provide a fork lift truck and driver to assist with the offloading and positioning of the equipment.

The vendor's personnel will work strictly in accordance with Contractors' Site rules and Health and Safety rules in operation at the customer. It is the responsibility of the Project Manager to ensure all personnel working on or associated with the project are made aware of the necessary safety requirements. The vendor will be responsible for providing suitable PPE for its personnel and subcontractors.

A.3.11 Final Testing and Buy-off

Once the system has been commissioned, the Site Acceptance Test (SAT) will be carried out to confirm compliance with the specification. The equipment will be formally taken over when the SAT has been successfully completed.

SAT trials are expected to take place within two working weeks of the completion of commissioning at a date to be agreed upon by both parties. If the vendor is prevented from conducting these trials within this period, through no fault of its own, then take-over will be considered to have taken place, and the customer will accept the equipment by default.

The vendor will provide standby cover with one engineer to remain on site for 5 working days following the completion of initial SAT. The engineer will not be required to operate the system but is to offer assistance in the event of problems or breakdowns.

A.3.12 SAT Procedure

The initial SAT will include a preliminary inspection of the equipment and documentation in line with the User Requirement Specification. This will be followed by functional testing as defined in the Functional Test Specification (see Section A.3.7). This will be similar to the FAT but will produce complete batches of each assembly variant (see Section A.2.1).

During these tests, the following will be recorded for each variant of assembly produced:

- Total operating time (TOT)
- The length and cause of any down time (DT)
- The time required to perform the fixture changeover (FCT).

The customer will provide personnel to load and unload the system under the supervision of the vendor.

Cycle Time and Availability Calculations

The total operating time recorded for the production of each variant will be compared with the cycle time estimates provided by the vendors for those assemblies at 100% efficiency (see Section A.2.4) to determine if the cycle time estimates have been achieved.

That is

$$\text{Cycle time} = \text{TOT}/\text{Batch size}$$

The causes of the down time will be reviewed to determine if they are due to the function of the equipment or other issues outside the performance of the cell. Any revisions will be incorporated with the down time (DT) figure.

The down time (DT) for each variant will then be summed to give a total down time figure (DT^{TOT}). Similarly, the fixture changeover time (FCT) between variants will be summed to give a total fixture changeover time (FCT^{TOT}), and the Total operating time (TOT) for the variants will be summed to give an overall total operating time (TOT^{TOT}).

The availability will then be determined:

$$\text{Availability} = TOT^{TOT} / \left(TOT^{TOT} + DT^{TOT} + FCT^{TOT} \right)$$

The availability will then be compared with the target specified in Section A.2.4.

If the equipment achieves the initial SAT successfully, the final SAT will then be performed to determine the reliability of the system. This will be performed during the 5 days of standby cover. The system will be operated, by the customer, under normal production conditions, and details of any down time will be recorded. The availability for this period will then be determined using the calculation shown above and compared with the requirement defined in Section A.2.4.

Assuming the equipment passes both the initial and final SAT, the system will be accepted by the customer. If the equipment fails at any stage, the vendor will be responsible for any rectification work, and the full SAT will recommence. The vendor is to continue with the standby cover until the SAT has been successfully completed.

A.3.13 Documentation

All equipment should be designed to be easily maintainable.

An 'as built' documentation package will be required, including:

- CE marking
- Electrical drawings
- Complete software listing including comments
- FDS describing the operation of the system
- System certifications
- AutoCAD drawings
- Full list of commercially available bought-in items together with their source and the reference number recognised by the source.

An Operation and Maintenance Manual will be supplied with the system. This will contain instructions for safe use, as required by the EEC directives and Machinery regulations. This will also detail fault recovery procedures and specify the recommended preventative maintenance including the procedures and required frequency. Details of recommended spares and consumable items are to be included in the manual. The Operation and Maintenance Manual will be supplied in the form of a 1 off CD.

A.3.14 Training

The customer fully recognises the importance of appropriate training tailored to its specific requirements. Our experience shows that a smooth handover of equipment can only be achieved if all users and supervisory staff are properly trained.

The proposal should include appropriate training, to take place before the SAT. This is to include training for four operators and two technicians. The technicians are to receive training on robot programming and maintenance, as well as training dedicated to the cell, including preventative maintenance, fault finding and error recovery. The proposal should detail the training to be provided.

Hands-on training is to continue during the one week of standby cover.

A.3.15 Spares and Service Contract

The proposal must identify an appropriate level of spare parts recommended for purchase. The final list of spare parts will be detailed by the Project Manager and the customer once the installation has been completed. Wear parts must also be highlighted, and the proposal is to include the supply of sufficient parts to operate the machine for 6 months on a two-shift basis.

The proposal is to include the offer of an appropriate service contract. This must include, at a minimum, annual preventative maintenance but may also include other options such as breakdown callout.

A.4 GENERAL

The following section provides general information relating to this project.

A.4.1 Contacts

The main point of contact at the customer for this project will be:

Name

Job title

Address
Telephone number
Mobile number
Email address

A.4.2 Clarifications

Any request for clarifications should be made to the contact identified in Section A.4.1. All clarifications must be confirmed in writing, with email being acceptable.

A.4.3 Environment

The environment is typical for a fabrication facility. The vendor must check the facility and proposed location for the system and identify any concerns it may have early in the project.

Although the system will be situated in an enclosed building, it may be exposed to temperatures below 0 °C when the facility is not working.

A.4.4 Preferred Vendors

The customer uses equipment from the following preferred Vendors:

- Pneumatics – XX
- Electrical – XX
- PLCs – XX
- Motor and Gearboxes – XX
- Welding equipment – XX

A.4.5 Warranty

The warranty period is to be 1 year from the date of successful completion of the SAT. The warranty is to include the replacement of all worn or damaged parts and all the time required to fit and restart the system including travel time and subsistence (excepting where neglect or misuse has occurred). Wear parts previously identified by the vendor (see Section A.3.15) are excluded from this warranty unless unacceptably fast wear has occurred. In this case the vendor will be required to investigate the cause and provide remedial action under the terms of the warranty.

A.4.6 Standards

The system and equipment must meet CE requirements conforming to all the relevant standards, including the Machinery Directive 2006/42/EC, Low Voltage Directive 2006/95/EC, and EMC Directive 2004/108/EC.

INDEX

Note: Page numbers followed by *f* indicate figures.

A

I

J

K

L

M

Y

Printed in the United States
By Bookmasters